2 View near the City of Lin-chhing on the banks of the Grand Canal, an engraving from Staunton's account of the Macartney Embassy of A.D. 1793. Lin-chhing was the hub of the system as it took its final form under the Yuan (Mongolian) Dynasty towards the end of the 13th century, embodying the oldest successful summit canal in any civilization. A European parallel would be a broad artificial river from London to Athens.

3 A bronze 'rainbow vessel' (*kung têng*) of the Han period (*c.* first century B.C. or A.D.), probably used for sublimation, front view. Cf. p. 70.

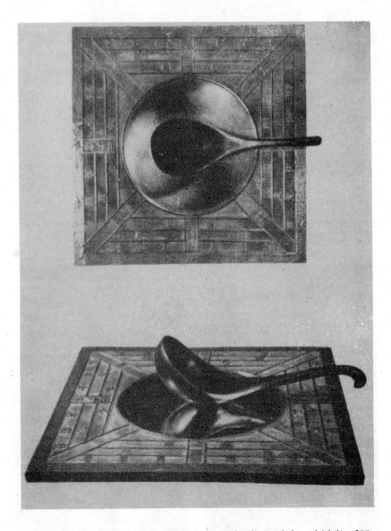

4 The earliest form of the magnetic compass; the diviner's board (*shih*) of Han times (first century B.C. or A.D.) with the lodestone spoon (*shao*) upon it (Wang Chen-To). Cf. p. 73.

5 Kuo Shou-ching's 'equatorial torquetum' (Simplified Instrument, *chien i*) of A.D. 1276, the first equatorial mounting; in the position it now occupies at the Purple Mountain Observatory near Nanking (original photo. 1958). Cf. p. 79.

6 The oldest certain evidence of foot-stirrups; tomb statuettes of the Chin Dynasty (*c.* A.D. 300) from Chhangsha (Kao Chih-Hsi *et al.*). Cf. p. 86.

7 An early example of collar harness; detail copy painting of the lower right cart in the fresco of the Procession of the Exarch's Consort at the Tunhuang cave-temples. (Cave 156, painted A.D. 851; copy painting by Ho Yi). Cf. p. 89.

8 Flour-mill on a mountain river; scroll-painting by an unknown artist of the Yuan Dynasty, *c.* A.D. 1300. In the tradition of all Chinese painters this artist worked not from the life, but in tranquil recollection; hence, not being a mill-wright, he confused paddle-wheels with gear-wheels. Nevertheless it is clear that a number of different machines were powered in this mill by two large horizontal water-wheels (centre and right lower compartments). The left upper compartment shows right-angle gearing, probably working a battery of trip-hammers. The centre upper compartment has the main mill-stones and an edge-runner mill in front of the staircase. The right upper compartment has a curious contraption almost certainly to be interpreted as an attempt to draw from memory the crank, connecting-rod and piston-rod combination, i.e. the water-powered reciprocator (cf. Fig. 13), working a flour-sifter, perhaps the latticed cupboard seen at the back of it. The left and centre lower compartments show a number of badly drawn gear-wheels, both horizontally and vertically mounted, the exact purpose and connection of which is not clear; but in the right lower compartment the artist has drawn a gear-wheel of tub shape, equally isolated but with admirably designed short pinion teeth, thus revealing to us the well-developed technique of the Yuan millwrights.

9 Han Dynasty (first century B.C. or A.D.) tomb model showing rotary mill, pedal tilt-hammer, and a rotary winnowing-fan worked with a crank handle; the oldest representation of a crank in any civilization. Cf. p. 95.

10 One of the oldest representations of a spinning-wheel in any culture, and hence important evidence on the history of the driving-belt; *A Son taking leave of his Mother*, probably by Chhien Hsüan, in any case datable *c.* A.D. 1270 (Waley). Cf. p. 100.

11 The oldest extant cast-iron pagoda (A.D. 1061), at Yü-Chhüan Ssu, Tang-yang, in Hupei province (Boerschmann). Cf. p. 103.

12 The earliest segmental arch bridge in any civilization, Li Chhun's An-chi Bridge at Chao-hsien, Hopei province, built in A.D. 610 (Mao I-Shêng). Cf. p. 105.

13 Han pottery model of a ship, from a first-century A.D. tomb at Canton; the oldest known example of the stern-post rudder in any culture. Cf. p. 109.

14 Detail of stern of **Han ship model** showing the **rudder.**

15 The old library pavilion at the **Confucian Temple** (Wên Miao) at Chiating, Szechuan province, in 1943, then the home of evacuated Wuhan University. I often visited there at the time paper 3 was written. It reminds us of the high literary culture associated with the numinous ethics of Confucianism all through the ages. (Orig. photo.)

16 The story of Mencius and his mother. Two pages from the *Lieh Nü Chuan* (Biographies of Notable Women) a book originating probably in the Han period (2nd century B.C. to 2nd century A.D.). Facsimile edition of 1820 copying a Ming print of 15th or 16th century, but the iconographic tradition, the style of picture, etc. goes back to the edition of A.D. 1063. The labels *in* the picture itself say (left) 'School' (or College); (right) 'Mencius' mother'. The loom, though badly drawn, is of the Han type for silk, the 45° or oblique–warp construction, a sign of antiquity. Mencius, i.e. Mêng Kho, 374 to 289 B.C., was one of the greatest Confucian scholars of all time. His name has always been a household word in China. The text reads as follows: 'The mother of Mêng Kho (Mencius) of Tsou was just called "Mêng's Mother". They lived in a cottage near some tombs. When Mencius was small he was fond of wandering about watching the funerals and the tomb-builders at work, so his mother said "This is not at all a good place to live", and moved to a house beside the market. But there little Mencius spent his time playing with the merchants and watching their trade. So she said again "This is not where we ought to be", and moved to a home beside a College. Here he enjoyed looking at the ceremonies of scholars at receptions and lectures, going and coming. Mêng's Mother said "This is the right place for my son."

When Mencius was young, he came back one evening from school when his mother was weaving on her loom. She said "Have you now learnt everything?" He replied "Yes, I think now I know all I need to". So she took her scissors and ripped right across the cloth, at which Mencius was frightened and asked her why she had done so. His mother answered "Your stupidity, not realising how much there is to learn, is like my cutting through the web that I was weaving, before it was finished. Men only become famous for their learning and intelligence by using all possible effort and hard work to acquire knowledge. People who ask them questions find how broad their knowledge is, they live peaceably in their habitations, and keep all harmful things far away. Now if you do not appreciate the heights of knowledge, this will infallibly invite calamity. How is that different from my weaving for a living, yet stopping halfway through the task? How would this clothe a man, and how could you, when grown up, not be short of food? An idle woman will suffer ill fare, and a boy who does not study hard will fail to make use of his capacities. How could he not end by becoming a thief or a robber, and at the last a prisoner or a slave?" Mencius was exceedingly frightened, and thenceforward studied day and night without resting.

When he grew older he became a master of the six arts, and one of the most famous scholars under heaven. He was the greatest of all the disciples of Confucius and Tzu-Ssu. People said that Mencius' mother had been wise in knowing what would influence her son. As the *Book of Odes* says:

> "Such a great gentleman,
> How shall we feed him?
> Such a great gentleman,
> What can we give to him?".'

17 'Backing the book' in traditional China; a student repeating a classical text
learnt by heart, in the presence of his teacher, (photo. Mencarini, in Bishop, *c.*
1895). When passing country schools as late as the Second World War I often
used to hear the hum of classical repetitions wafting through the open windows.
Cf. p. 179.

18 The old examination cells at Nanking, (photo.Williams, in Hutchinson, *c.* 1925). Under the system of the imperial examinations aspirants who had won their bachelor's degrees locally used to assemble at the provincial capital for the examinations for the master's degree. Each candidate spent his days in one of a vast range of cells like riverside changing-cabins, guarded by invigilators and supplied with frugal meals brought in from outside. In the late Chhing period, as in the Ming, the subjects were confined to orthodox literature and philosophy, a particular stilted style being expected; but in earlier times the examinations had had much more to do with concrete administrative, governmental and economic problems, while in some dynasties such as the Thang and Sung, technical subjects such as astronomy, engineering and medicine could be taken. Cf. p. 179.

19 Chinese students of today and yesterday; undergraduates at Tung-chi University, then evacuated to Lichuang in Szechuan province (1943), and occupying the Ta Yü Temple where I took the picture after one of my own lectures.

20 Another part of the Grand Canal (cf. Pl. 2), also an engraving from Staunton's account of the Macartney Embassy of 1793. The canal at one time ran through Pao-Ying lake in Chiangsu, but for the protection of traffic a dyke was erected along this section in A.D. 1007 separating the canal from the open lake. Cf. p 195.

21 A *diolkos* (double slipway) on the Grand Canal; a drawing by Allom & Wright, 1843. This device, which goes far back into antiquity, both Chinese and Greek, permitted vessels of considerable burden to be transferred between canal stretches of different water-levels, using massive double capstans. The drawing, though attractive, exaggerates the slipway angle and romanticises the Chinoiserie. The pound-lock, however, was a purely Chinese invention, dating from A.D. 984, when ChhiaoWei-Yo constructed the first such installation. Cf. p. 194.

22 At the Confucian temple (Wên Miao) at Chhêng-kung south of Kunming, 1942, then the home of the Institute of Population Statistics of the United Universities evacuated to Yunnan province. In the background, the main shrine with the tablet of the Sage, to the right one can see some of the tablets of his disciples. On the altar of the Sage stands a calculating-machine, which I thought most appropriate for one who had been so conscious of the welfare of the people—'Feed them, educate them' comes as a refrain in the *Lun Yü* (Conversations and Discourses), 5th century B.C. Population records and problems have been a favourite study of Chinese scholars for centuries, and since the foundation of the People's Republic there has been constant attention to the means and spread of family planning. As for refusal of inventions because of fear of technological unemployment, one can point to hardly any case of this in China during the past two thousand years. Cf. p. 208.

THE GRAND TITRATION

THE GRAND TITRATION

SCIENCE AND SOCIETY IN EAST AND WEST

Joseph Needham F.R.S.

London
GEORGE ALLEN & UNWIN LTD
RUSKIN HOUSE   MUSEUM STREET

FIRST PUBLISHED IN 1969
SECOND IMPRESSION 1972

© *George Allen & Unwin Ltd,* 1969

ISBN 0 04 931005 4

PRINTED IN GREAT BRITAIN
in 12 on 13 point Bembo type
BY LOWE AND BRYDONE (PRINTERS) LTD
LONDON

To

LU GWEI-DJEN

the explainer, the antithesis,
the manifestation,
the assurance of a link
no separation can break

The system of romanization of Chinese names used in this book is that of Wade-Giles with *h* substituted for the aspirate apostrophe.

CONTENTS

INTRODUCTION

The historical civilization of China is, with the Indian and the European-Semitic, one of the three greatest in the world, yet only in recent years has any enquiry been begun into its contributions to science and technology. Apart from the great ideas and systems of the Greeks, between the first and the fifteenth centuries the Chinese, who experienced no 'dark ages', were generally much in advance of Europe; and not until the scientific revolution of the late Renaissance did Europe draw rapidly ahead. Before that time, however, the West had been profoundly affected not only in its technical processes but in its very social structures and changes by discoveries and inventions emanating from China and East Asia. Not only the three which Lord Bacon listed (printing, gunpowder and the magnetic compass) but a hundred others—mechanical clockwork, the casting of iron, stirrups and efficient horse-harness, the Cardan suspension and the Pascal triangle, segmental-arch bridges and pound-locks on canals, the stern-post rudder, fore-and-aft sailing, quantitative cartography—all had their effects, sometimes earth-shaking effects, upon a Europe more socially unstable.

Why, then, did modern science, as opposed to ancient and medi-eval science (with all that modern science implied in terms of political dominance), develop only in the Western world? Nothing but a careful analysis, a veritable titration, of the cultures of East and West will eventually answer this question. Doubtless many factors of an intellectual and philosophical character played their part, but there were certainly also important social and economic causes which demand investigation.

Now that everyone learns elementary chemistry at school, one need not fear to speak of titration. Most people have handled a burette, and seen the colour change when the reaction is completed. The lexicographers have not been very successful with their definitions in the usual dictionaries, but we may say that titration is the determination of the quantity of a given chemical compound in a solution by observing the amount of a solution of another compound at known strength required to convert the first completely into a third, the end-point being ascertained by a change of colour or other means. This is what is called volumetric analysis, and it is more recent than one would think. A precursor process was used by Guyton de Morveau in 1782, but the technique was systematized fully by John Dalton and described in a paper of 1819. Still the name did not come, however, for 1864 saw the first use of the word—derived undoubtedly from the French *titre*, an assayists' term used much earlier for the degree of purity of gold in alloys. Now in my work with my collaborators on the history of discovery and invention in Chinese and other cultures we are always trying to fix dates—the first canal lock in China in A.D. 984, the first irrigation contour canal in Assyria in 690 B.C., the first transport contour canal in China in 219 B.C., the first eye-glasses or spectacles in Italy in A.D. 1286, and so on. In such a way can one 'titrate' the great civilizations against one another, to find out and give credit where credit is due, and so also, it seems, must one analyse the various constituents, social or intellectual, of the great civilizations, to see why one combination could far excel in medieval times while another could catch up later on and bring modern science itself into existence.

Hence the title of this book of papers, lectures and essays written on various occasions, paralleling but not overlapping our volumes of *Science and Civilisation in China*.[1] Gathered together upon the suggestion of my friend Ruth Nanda Anshen, I hope that they will throw some interim light upon this great and paradoxical theme in the comparative sociology of knowledge. Lastly it only remains to offer thanks to my collaborators from Cathay and

---

[1] J. Needham, with the collaboration of Wang Ling (Wang Ching-Ning), Lu Gwei-Djen, Ho Ping-Yü, Kenneth Robinson, Tshao Thien-Chhin and others: *Science and Civilisation in China* (seven volumes in twelve parts), Cambridge, 1954– ; hereinafter abbreviated as *SCC*.

Manzi for thirty years of laborious excavations in daily company—
Wang Ching-Ning, Lu Gwei-Djen, Tshao Thien-Chhin, Ho
Ping-Yü—without whom nothing could have been accomplished
in piercing the barrier between the ideographic and the alphabetic
civilizations; and to my wife, whose wise assay clarifies all that is
written.

J.N.

POVERTIES AND TRIUMPHS OF THE CHINESE SCIENTIFIC TRADITION

First published in *Scientific Change* (Report of History of Science Symposium, Oxford, 1961), ed. A. C. Crombie, London 1963

INTRODUCTION

In what follows an attempt will be made to describe some of the elements of strength and weakness in the growth and development of the indigenous Chinese tradition of science and invention, in contrast with that of Europe. The inspiration of the title comes, of course, from certain famous phrases coined by French authors of the last century. I refer to the '*Servitudes et grandeurs*' of the military life about which Alfred de Vigny wrote, and later on the '*Splendeurs et misères*' of courtesans immortalized by Honoré de Balzac. Both East and West had strengths and weaknesses now well discernible as we look back along the course which man's knowledge of nature and control of nature took in the diverse regions of the Old World. My object will be to describe some of the outstanding contrasts between the Chinese and European traditions in the natural sciences, pure and applied, then to say something about the position of scientists and engineers in classical Chinese society, and lastly to discuss certain aspects of science in relation to philosophy, religion, law, language, and the concrete circumstances of production and exchange of commodities.

First of all it is essential to define the differences between ancient and medieval science on the one hand, and modern science on the other. I make an important distinction between the two. When we say that modern science developed only in Western

POVERTIES AND TRIUMPHS

Europe at the time of Galileo in the late Renaissance, we mean surely that there and then alone there developed the fundamental bases of the structure of the natural sciences as we have them today, namely the application of mathematical hypotheses to Nature, the full understanding and use of the experimental method, the distinction between primary and secondary qualities, the geometrisation of space, and the acceptance of the mechanical model of reality. Hypotheses of primitive or medieval type distinguish themselves quite clearly from those of modern type. Their intrinsic and essential vagueness always made them incapable of proof or disproof, and they were prone to combine in fanciful systems of gnostic correlation. In so far as numerical figures entered into them, numbers were manipulated in forms of 'numerology' or number-mysticism constructed *a priori*, not employed as the stuff of quantitative measurements compared *a postiori*. We know the primitive and medieval Western scientific theories, the four Aristotelian elements, the four Galenical humours, the doctrines of pneumatic physiology and pathology, the sympathies and antipathies of Alexandrian proto-chemistry, the *tria prima* of the alchemists, and the natural philosophies of the Kabbala. We tend to know less well the corresponding theories of other civilizations, for instance the Chinese theory of the two fundamental forces Yin and Yang, or that of the five elements, or the elaborate system of the symbolic correlations. In the West Leonardo da Vinci, with all his brilliant inventive genius, still inhabited this world; Galileo broke through its walls. This is why it has been said that Chinese science and technology remained until late times essentially Vincian, and that the Galilean break-through occurred only in the West. That is the first of our starting-points.

Until it had been universalized by its fusion with mathematics, natural science could not be the common property of all mankind. The sciences of the medieval world were tied closely to the ethnic environments in which they had arisen, and it was very difficult, if not impossible, for the people of those different cultures to find any common basis of discourse. That did not mean that inventions of profound sociological importance could not diffuse freely from

15

one civilization to another—mostly in fact from east to west. But the mutual incomprehensibility of the ethnically-bound concept systems did severely restrict possible contacts and transmissions in the realm of scientific ideas. This is why technological elements spread widely through the length and breadth of the Old World, while scientific elements for the most part failed to do so.

Nevertheless the different civilizations did have scientific interchanges of great importance. It is surely quite clear by now that in the history of science and technology the Old World must be thought of as a whole. Even Africa may have been within its circuit. But when this oecumenical view is taken, a great paradox presents itself. Why did modern science, the mathematization of hypotheses about Nature, with all its implications for advanced technology, take its meteoric rise *only* in the West at the time of Galileo? This is the most obvious question which many have asked but few have answered. Yet there is another which is of quite equal importance. Why was it that between the second century B.C. and the sixteenth century A.D. East Asian culture was much *more* efficient than the European West in applying human knowledge of Nature to useful purposes? Only an analysis of the social and economic structures of Eastern and Western cultures, not forgetting the great role of systems of ideas, will in the end suggest an explanation of both these things.

THE FACE OF SCIENCE AND TECHNOLOGY IN TRADITIONAL CHINA

Before the river of Chinese science flowed, like all other such rivers, into this sea of modern science, there had been remarkable achievements in mathematics. Decimal place-value and a blank space for the zero had begun in the land of the Yellow River earlier than anywhere else, and decimal metrology had gone along with it. By the first century B.C. Chinese artisans were checking their work with sliding calipers decimally graduated. Chinese mathematical thought was always profoundly algebraic, not geometrical, and in the Sung and the Yuan (twelfth to fourteenth centuries A.D.) the Chinese school led the world in the solution of equations, so that the triangle called by Pascal's name was already old in China in A.D. 1300. We often find examples of this sort; the system of linked and pivoted rings which we know as the Cardan suspension was commonly used in China a thousand years before Cardan's time. As for astronomy, I need only say that the Chinese were the most persistent and accurate observers of celestial phenomena anywhere before the Renaissance. Although geometrical planetary theory did not develop among them they conceived an enlightened cosmology, mapped the heavens using our modern co-ordinates, and kept records of eclipses, comets, novae and meteors still useful, for example to the radio-astronomers, today. A brilliant development of astronomical instruments also occurred, including the invention of the equatorial mounting and the clock-drive; and this development was in close dependence upon the contemporary capabilities of the Chinese engineers. Their skill affected also other sciences such as seismology, for it was a Chinese man of science, Chang Hêng, who built the first practical seismograph about A.D. 130.

Three branches of physics were particularly well developed in ancient and medieval China—optics, acoustics and magnetism. This was in striking contrast with the West where mechanics and dynamics were relatively advanced but magnetic phenomena almost unknown. Yet China and Europe differed most profoundly perhaps in the great debate between continuity and discontinuity, for just as Chinese mathematics was always algebraic rather than geometrical, so Chinese physics was faithful to a prototypic wave theory and perennially averse to atoms. One can even trace such contrasts in preferences in the field of engineering, for whenever an engineer in classical China could mount a wheel horizontally he would do so, while our forefathers preferred vertical mountings—water-mills and wind-mills are typical examples.

A pattern which we very often find in comparing China's achievements with those of Europe is that while the Chinese of the Chou, Chhin and Han, contemporary with the Greeks, did not rise to such heights as they, nevertheless in later centuries there was

17

nothing in China which corresponded to the period of the Dark Ages in Europe. This shows itself rather markedly in the sciences of geography and cartography. Although the Chinese knew of discoidal cosmographic world-maps, they were never dominated by them. Quantitative cartography began in China with Chang Hêng and Phei Hsiu about the time when Ptolemy's work was falling into oblivion, indeed soon after his death, but it continued steadily with a consistent use of the rectangular grid right down to the coming of the Jesuits in the seventeenth century A.D. The Chinese were also very early in the field with advanced survey methods and the making of relief maps. In the geological sciences and in meteorology the same pattern presents itself.

Mechanical engineering and indeed engineering in general was a field in which classical Chinese culture scored special triumphs. Both the forms of efficient harness for equine animals—a problem of linkwork—originated in the Chinese culture-area, and there also water-power was first used for industry about the same time as in the West (first century A.D.); not, however, so much for grinding cereals as for the operation of metallurgical bellows. The development of iron and steel technology in China constitutes a veritable epic, with the mastery of iron-casting some fifteen centuries before its achievement in Europe. Contrary to the usual ideas, mechanical clockwork began not in early Renaissance Europe but in Thang China, in spite of the highly agrarian character of East Asian civilization. Civil engineering also shows many extraordinary achievements, notably iron-chain suspension bridges (Fig. 1) and the first of all segmental arch structures (Pl. 12), the magnificent bridge built by Li Chhun in A.D. 610. Hydraulic engineering was always prominent in China on account of the necessity of control of waterways for river conservation (defence against flood and drought), irrigation, and tax-grain transport.

In martial technology the Chinese also showed notable inventiveness. The first appearance of gunpowder occurs among them in the ninth century A.D., and from A.D. 1000 onwards there was a vigorous development of explosive weapons some three centuries before they appeared in Europe. Probably the key in-

FIGURE 1   A Chinese picture of a suspension bridge. Although such bridges of wrought-iron chains go back to the Sui (6th century A.D.) at least, long anticipating European designs (16th century A.D.) and first realizations (18th century A.D.), pictures of them in traditional style are very rare. This one is taken from the *Mei-Shan Thu Chih* (Illustrated Record of Mt O-mei in Szechuan).   The bridge is but a small one compared with the heavy-duty structures quite common in the western provinces. It is labelled 'Bridge of No Regrets (for a Worldly Life),' and three people are passing across it on their pilgrimage to a Buddhist temple higher up.

vention was that of the fire-lance at the beginning of the twelfth
century A.D., in which a rocket composition enclosed in a bamboo
tube was used as a close-combat weapon. From this derived, I
have little doubt, all subsequent barrel guns and cannon of whatever material constructed. Other aspects of technology also have
their importance, especially that of silk in which the Chinese
excelled so early. Here the mastery of a textile fibre of extremely
long staple appears to have led to the first development of technical
devices so important as the driving-belt and the chain-drive. It is
also possible to show that the first appearance of the standard
method of converting rotary to longitudinal motion is found in
connexion with later forms of the metallurgical blowing-engine
referred to above. I must pass over other well-known inventions
such as the development of paper, block-printing and movable-
type printing, or the astonishing story of porcelain.

There was no backwardness in the biological fields either, and
here we find many agricultural inventions arising from an early
time. As in other subjects, we have texts which parallel those of
the Romans such as Varro and Columella from a similar period.
If space permitted, one could take examples from plant protection
which would include the earliest known use of the biological control of insect pests. Medicine is a field which aroused the interests
of the Chinese in all ages, and which was developed by their
special genius along lines perhaps more different from those of
Europe than in any other case. I think that I can do no more here
than refer simply to one remarkable fact, namely that the Chinese
were free from the prejudice against mineral remedies which was
so striking in the West; they needed no Paracelsus to awaken them
from their Galenical slumbers for in these they had never participated. They were also the greatest pioneers of the techniques of
inoculation.

CONTRASTS BETWEEN CHINA AND THE WEST

Let us come now to the further examination of some of the great
contrasts to which I have already referred. In the first place it can

be shown in great detail that the *philosophia pere.inis* of China was an organic materialism. This can be illustrated from the pronouncements of philosophers and scientific thinkers of every epoch. The mechanical view of the world simply did not develop in Chinese thought, and the organicist view in which every phenomenon was connected with every other according to hierarchical order was universal among Chinese thinkers. Nevertheless this did not prevent the appearance of great scientific inventions such as the seismograph, to which we have already referred. In some respects this philosophy of Nature may even have helped. It was not so strange or surprising that the lodestone should point to the pole if one was already convinced that there was an organic pattern in the cosmos. If, as is truly the case, the Chinese were worrying about the magnetic declination before Europeans even knew of the polarity, that was perhaps because they were untroubled by the idea that for action to occur it was necessary for one discrete object to have an impact upon another; in other words, they were inclined *a priori* to field theories, and this predilection may very well also account for the fact that they arrived so early at a correct conception of the cause of sea tides. One may find remarkable statements, as early as the San Kuo period, of action at a distance taking place without any physical contact across vast distances of space.

Again, as we have said, Chinese mathematical thought and practice was invariably algebraic, not geometrical. No Euclidean geometry spontaneously developed among them, and this was doubtless inhibitory for the advances they were able to make in optics, where however, incidentally, they were never handicapped by the rather absurd Greek idea that rays were sent forth by the eye. Euclidean geometry was probably brought to China in the Yuan (Mongol) period but did not take root until the arrival of the Jesuits. Nevertheless all this did not prevent the successful realization of great engineering inventions—we have mentioned two already, the most useful method of interconversion of rotary and rectilinear motion by means of eccentric, connecting-rod and piston-rod; and the successful achievement of the oldest form of mechanical clock. What this involved was the invention of an escapement, namely a mechanical means of slowing down the revolution of a set of wheels so that it would keep time with humanity's primary clock, the apparent diurnal revolution of the heavens. In this connexion it is interesting to find that Chinese practice was not, as might at first sight be supposed, purely empirical. The successful erection of the great clock-tower of Su Sung at Khaifêng in A.D. 1088 was preceded by the elaboration of a special theoretical treatise by his assistant Han Kung-Lien, which worked out the trains of gears and general mechanism from first principles. Something of the same kind had been done on the occasion of the first invention of this kind of clock by I-Hsing and Liang Ling-Tsan early in the eighth century A.D., six centuries before the first European mechanical clocks with their verge-and-foliot escapements. Moreover, though China had no Euclid, that did not prevent the Chinese from developing and consistently employing the astronomical co-ordinates which have completely conquered modern astronomy and are universally used today, nor did it prevent their consequent elaboration of the equatorial mounting, although there was nothing but a sighting-tube, and as yet no telescope, to put in it.

Thirdly, there is the wave-particle antithesis. The prototypic wave theory with which the Chinese concerned themselves from the Chhin and Han onwards was connected with the eternal rise and fall of the two basic natural principles, the Yang and Yin. From the second century A.D. onwards atomistic theories were introduced to China time after time, especially by means of the Buddhist contacts with India, but they never took any root in Chinese scientific culture. All the same this lack of particulate theory did not prevent the Chinese from curious achievements such as the recognition of the hexagonal system of snowflake crystals many centuries before this was noticed in the West. Nor did it hinder them from helping to lay the foundation of knowledge of chemical affinity, as was done in some of the alchemical tractates of the Thang, Sung and Yuan. There the absence of particulate conceptions was probably less inhibitory than it otherwise might have been, because it was only after all in the post-Renaissance period in Europe that these theories became so fundamental for the rise of modern chemistry.

I should not want to disagree altogether with the idea that the Chinese were a fundamentally practical people, inclined to distrust all theories. One must beware, however, of carrying this too far, because the Neo-Confucian school in the eleventh, twelfth and thirteenth centuries A.D. achieved a wonderful philosophical synthesis strangely parallel in time with the scholastic synthesis of Europe. One might also say that the disinclination of the Chinese to engage in theory, especially geometrical theory, brought advantages with it. For example, Chinese astronomers did not reason about the heavens like Eudoxus or Ptolemy but they did avoid the conception of crystalline celestial spheres which dominated medieval Europe. By a strange paradox, when Matteo Ricci came to China at the end of the sixteenth century A.D. he mentioned in one of his letters a number of the foolish ideas entertained by the Chinese, among which prominently figured the fact that 'they do not believe in crystalline celestial spheres'; it was not long before the Europeans did not either. Moreover, this fundamental practicality did not imply an easily satisfied mind. Very careful experimentation was practised in classical Chinese culture. For example the discovery of magnetic declination would not have occurred unless the geomancers had been attending most carefully to the positions of their needles, and the triumphs of the ceramics industry could never have been achieved without fairly accurate temperature measurement and the means of repetition at will of oxidizing or reducing conditions within the kilns. The fact that relatively little written material concerning these technical details has come down to us springs from social factors which prevented the publication of the records which the higher artisans certainly kept. Enough remains, either by title, like the *Mu Ching* (Timberwork Manual) which we shall speak of again, or in MS. form, like the Fukien shipwrights' manual, to show that this literature existed.

THE SOCIAL POSITION OF SCIENTISTS AND
ENGINEERS IN TRADITIONAL CHINA

Something must now be said of the social position of scientists, engineers and artisans in Chinese feudal-bureaucratic society. The first factor which springs to mind is the relatively 'official' character of science, pure and applied. The astronomer, as has been well said, was not a citizen on the outskirts of the conventions of his society, as perhaps in the Greek city-states, but a civil servant lodged at times in part of the imperial palace, and belonging to a bureau that was an integral part of the civil service. On a lower intellectual plane, no doubt, the artisans and engineers also participated in this bureaucratic character, partly because in nearly all dynasties there were elaborate imperial workshops and arsenals, and partly because during certain periods at least those trades which possessed the most advanced techniques were 'nationalized', as in the Salt and Iron Authority under the Former Han. One finds also a strong tendency for technicians to gather round the figure of one or another prominent official who encouraged them and supported them as his personal followers. At the same time there is no possible doubt that throughout the ages there was always a large realm of handicraft production independently undertaken by and for the common people. No doubt when any large, new or unusually complex piece of machinery was constructed (such as the early water-mills and the early mechanical clocks), or any outstandingly large civil engineering project undertaken, all this was done either in the imperial workshops or under the close supervision of important officials. The imperial workshops went by many names of which Shang-Fang was one of the most usual. Occasionally the names of some of the artisans working in these factories have come down to us. We have, for example, an inscription on the lid of a black lacquer box with a date equivalent to 4 B.C., which is remarkably interesting because it bears the names of no less than five administrators for seven technicians. We may conclude that something like 'Parkinson's Law' was already manifesting itself in ancient China.

The imperial workshops were situated not only at the capitals of successive dynasties but also in the most important provincial cities, the nodes in the administrative network. In the relatively private sector, particular localities derived fame from skills which tended to concentrate at sites of natural resources; one thinks of the lacquer-makers of Fuchow, the potters of Ching-tê-chen or the well-drillers (for brine and natural gas) of Tzu-liu-ching in Szechuan. However, Chinese technical skill tended to wander far and wide; there were Chinese metallurgists and well-drillers in second-century A.D. Parthia and Ferghana, while eighth-century A.D. Samarqand knew Chinese weavers and paper-makers. People were always asking for Chinese technicians; for example in A.D. 1126 when the Chin Tartars besieged the Sung capital at Khaifêng, all kinds of craftsmen were asked for as hostages, and as late as A.D. 1675 a Russian diplomatic mission officially requested that Chinese bridge-builders should be sent to Russia.

As regards the question of status, it is a very difficult one and still under investigation. The technical workers we have been mentioning were for the most part free plebeians (*shu-jen* or *liang-jen*). Only in very few cases do we hear of slaves or semi-servile people mentioned as producers of wealth; certain classical passages indeed specify free workers, as in the salt factories of the Han. Of course whatever the extent of government-organized production from time to time, the State relied upon an inexhaustible supply of obligatory unpaid labour in the form of the *corvée* (*yao* or *kung-yu*). In the Han period every male commoner between the ages of twenty and fifty-six was liable for one month's labour service a year; technical workers performed these obligations in the imperial workshops or at the nationalized factories, which were never primarily staffed by slaves. Eventually there grew up the practice of paying dues in lieu of personal service, so that a large body of artisans 'permanently on the job' (*chhang-shang*) resulted. Among the slave or semi-servile portion of the population there were no doubt a certain number of artisans, but it is highly doubtful whether it ever exceeded ten per cent in any period of Chinese history. The problem of servile and semi-servile status in Chinese

25

civilization is still very much under discussion, but most Western scholars believe that it was always primarily domestic in character and that the class was recruited by penal process; to become a convict meant to be 'enslaved to the State' for a term of years or for life, after which the prisoner could be allocated either to houses of great officials or to the imperial workshops or State factories.

Research now under way will throw much light upon the forms of slavery, semi-servility, free labour and the *corvée* system in classical Chinese society. Whatever the details of the conclusions which will be reached, we are constrained already to notice that ancient and medieval Chinese labour conditions proved no bar to a long series of 'labour-saving' inventions altogether prior to those arising in Europe and Islam. I shall return to this point a little later on. I want first to glance over some of the chief categories into which the lives of eminent scientists and engineers of ancient and medieval Chinese society seem to fall. One may divide such life-histories into five groups: first, high officials, the scholars who had successful and fruitful careers; second, commoners; third, members of the semi-servile groups; fourth, those who were actually enslaved; and fifth, a very significant group of minor officials, that is to say, scholars who were not able to make their way upwards in the ranks of the bureaucracy. The number of examples which we have found in our work varies very considerably among these different groups.

In the first place, among the high officials, we have already mentioned Chang Hêng. Chang Hêng was not only the inventor of the first seismograph in any civilization but the first to apply motive power to the rotation of astronomical instruments, one of the outstanding mathematicians of his time, and the father-figure in the design of armillary spheres. He became the President of the Imperial Chancellery. High provincial officials are often credited with important technical developments. Thus the introduction of the water-powered metallurgical blowing-engine is attributed to Tu Shih, who was Prefect of Nanyang in A.D. 31. Occasionally we find also a eunuch prominent in technical advance; the most obvious case is that of Tshai Lun, who began as a confidential secretary to the Emperor, was made Director of the Imperial Workshops in *paper* A.D. 97, and announced the invention of paper in A.D. 105.

On the contributions to science of Chinese princes and the remoter relatives of the imperial house an interesting monograph could be written. They were favoured with leisure, since though generally well educated they were in most dynasties ineligible for the civil service and yet disposed of considerable wealth. Though most of them did nothing for posterity, there were a memorable few who devoted time and riches to scientific pursuits. Liu An, the Prince of Huai-Nan (*fl.* 130 B.C.), with his entourage of naturalists, alchemists and astronomers, is one of the most famous figures in Chinese history. Another Han prince, Liu Chhung (*fl.* A.D. 173), is relevant also since he was the inventor of grid sights for crossbows and a famous shot with them into the bargain. In the Thang we meet with Li Kao, Prince of Tshao (*fl.* A.D. 784), interested in acoustics and physics but prominent here because of his successful development of treadmill-operated paddle-wheel warships.

Curiously enough, it seems exceptional to find an important engineer who attained high office in the Ministry of Works, at any rate before the Ming. This was probably because the real work was always done by illiterate or semi-illiterate artisans and mastercraftsmen who could never rise across that sharp gap which separated them from the 'white-collar' literati in the offices of the Ministry above. Still there were exceptions. There was Yuwên Khai (*fl.* A.D. 600), chief engineer of the Sui dynasty for thirty years. He carried out irrigation and conservation works, superintending the construction of what afterwards became part of the Grand Canal; he built large sailing carriages, and with Kêng Hsün devised the standard steelyard clepsydra of the Thang and Sung. One word more here about the tendency for technicians to cluster in the entourages of distinguished civil officials, who acted as their patrons. It is more than probable that the first water-mills and metallurgical blowing-engines were the work of technicians in the service of Tu Shih. Here an outstanding example is Shen Kua (*fl.* A.D. 1080), one of the greatest cientificminds in Chinese history, an ambassador, too, and an elder statesman. In his interesting and many-sided scientific book, the *Mêng Chhi Pi Than* (Dream Pool Essays), he describes the invention of movable-type printing by Pi Shêng about A.D. 1045, and says that after this commoner died, his fount of type "passed into the possession of my followers, among whom it has been kept as a precious possession until now". This gives a striking glimpse of the technicians gathered round important official patrons.

No doubt the greatest group of inventors is represented by commoners, master-craftsmen, artisans who were neither officials, even minor ones, nor of the semi-servile classes. Besides Pi Shêng, just mentioned, we have the great builder of pagodas, Yü Hao, who assuredly had to dictate his celebrated *Mu Ching* (Timberwork Manual) to a scribe. Yü Hao was a man of the tenth century A.D. but we may find his like in every dynasty. The second century A.D. saw the life of Ting Huan, renowned for his pioneer development of the Cardan suspension, the seventh century A.D. was the time of Li Chhun, the constructor of segmental arch bridges already mentioned, and the twelfth century A.D. the time of the greatest naval architect in Chinese history, Kao Hsüan, who specialized in the making of warships with multiple paddle-wheels. Sometimes we do not even have the surname—an omission which makes one wonder whether such men were not living on the borders of one of the semi-servile groups where surnames were not customary; for example there was the old craftsman (*lao kung*) who made astronomical apparatus in the first century B.C., or again that "artisan from Haichow" who presented to the Empress in A.D. 692 what was in all probability a complicated anaphoric clock. In the category of regular commoners we should probably also place minor military officers. Here I should like to mention Chhiwu Huai-Wên, a Taoist swordsmith who served in the army of Kao Huan the 'king-maker', founder of the Northern Chhi dynasty, and took charge of his arsenals (*c*. A.D. 545). Chhiwu Huai-Wên was one of the earliest protagonists, if not the inventor, of the co-fusion process of steelmaking, a process ancestral to the Siemens-Martin open-hearth furnace, and he was also a celebrated practitioner of the pattern-welding of swords.

And now we come to the exceptional cases, those who came down in history as brilliant scientific or technical men and yet whose social standing in their own time was very low indeed. The only one of clearly semi-servile rank in our registers is Hsintu Fang (*fl.* A.D. 525). In his youth he entered the household of a prince of the Northern Wei, Thopa Yen-Ming, as a 'dependent' or 'retainer'. This prince had collected many pieces of scientific apparatus—armillary spheres, celestial globes, trick hydrostatic vessels, seismographs, clepsydras, wind-gauges, etc., and had also inherited a very large library. As a client or pensioner of known scientific skill Hsintu Fang's position with relation to Thopa Yen-Ming must have been something like that of Thomas Hariot's to the ninth Earl of Northumberland in our own country (*c.* A.D. 1610). It seems that the prince intended to write certain scientific books with the aid of Hsintu, but owing to political and military events felt obliged to flee to the Liang emperor in the south in A.D. 528, so that Hsintu Fang had to write the books himself. After this he remained in seclusion, probably in poverty, until he was called to the court of another potentate, Mujung Pao-Lo, Governor of Tung-shan, whose younger brother recommended him to Kao Huan (the 'king-maker' just mentioned). This great lord he served as estate agent, a post which exercised his talents in surveying and architecture. He never rose higher than the position of dependent in one or other of the great semi-aristocratic houses but nevertheless left behind him a very high reputation in Chinese scientific history. Such an ingenious man of lowly origins could thus find shelter in a troublous time, if not official recognition or any considerable status, in the homes of unusual patricians.

Examples of technologists who were positively slaves are very rare indeed, but we have already mentioned Kêng Hsün (*fl.* A.D. 590). He began as a client of a Governor of Ling-nan, but when this patron died, Kêng Hsün, instead of going home, joined some tribal people in the south, and eventually led them in an uprising. When this was defeated and Kêng Hsün captured, a general, Wang Shih-Chi, realizing his technical ability, saved him from death and admitted him among his family slaves. Here his position was yet not so low that he could not receive instruction from an old friend, Kao Chih-Pao, who had in the meantime become Astronomer-Royal, and it was as a result of this that Kêng Hsün built an armillary sphere or celestial globe rotated continuously by water-power. The Emperor rewarded him for this by making him a government slave and attaching him to the Bureau of Astronomy and Calendar. The following Emperor freed him altogether, and he eventually rose to the position of Acting Executive Assistant in the Astronomical Bureau. His case shows that a long period of slavery was no bar to official, if not very exalted, positions.

We now reach the last of our group of technicians and one of the most numerous, namely that of the minor officials—men who were sufficiently well educated, even if of lowly origin, to enter the ranks of the bureaucracy, but whose particular talents or personalities frustrated all hopes of a brilliant career. Among these men we should include Li Chieh (*fl.* A.D. 1100), the man who, building on the earlier works of Yü Hao and others, produced the greatest definitive treatise of any age on the millennial tradition of Chinese architecture and building technology, the *Ying Tsao Fa Shih.* Li Chieh never rose above the rank of Director of Buildings and Construction and ended his official career as a magistrate in a provincial town. Yen Su (*fl.* A.D. 1030) was a Leonardo-like figure —scholar, painter, technologist and engineer under the Sung Emperor Jen Tsung. He designed the type of clepsydra with an overflow tank which remained standard for long afterwards, invented special locks and keys, and left specifications for hydro-static vessels, hodometers and south-pointing carriages. His writings included treatises on time-keeping and on the tides, but most of his life was spent in provincial administrative posts, and although he did become an Academician-in-waiting of the Lung-Thu Pavilion, he never rose above the position of Chief Executive Officer of the Ministry of Rites, and had no connexion with the Ministry of Works or other technical directorates. It was just the same with the two men of greatest practical importance in the pre-European history of the mechanical clock. Liang Ling-Tsan, the assistant of I-Hsing in the eighth century A.D., was a minor official in the War Office; and Han Kung-Lien, the principal collaborator of Su Sung 350 years later, was only an acting secretary in the Ministry of Personnel. It was this group of minor officials which provided one of the most striking texts on the life of technologists and scientists in medieval China which have come down to us. The engineer Ma Chün (fl. A.D. 260) was a man of outstanding ingenuity; he improved the drawloom, constructed a puppet theatre operated by water-power, invented the square-pallet chain-pump used so widely throughout the Chinese culture-area afterwards, designed (like Leonardo later on) a rotary arcu-ballista, and successfully constructed a south-pointing carriage, almost certainly making use of a simple form of differential gear. His friend Fu Hsüan devoted to his memory a remarkable essay—the text to which we have alluded. Fu Hsüan describes how Ma Chün was quite incapable of arguing with the sophisticated scholars nursed in the classical literary traditions, and in spite of all the efforts of his admirers, could never attain any position of importance in the service of the State, or even the means to prove by practical test the value of the inventions which he made. No document throws more light than this upon the inhibitory factors affecting science and technology which arose from the feudal-bureaucratic tradition of the scholar-gentry.

FEUDAL-BUREAUCRATIC SOCIETY

Next, one may ask what the long-term effects of the feudal-bureaucratic system, the mandarinate in fact, really were. In classical Chinese society certain sciences were orthodox and others the opposite. The institution of the calendar and its importance for a primarily agrarian society, and also to a lesser extent the belief in State astrology, made astronomy always one of the orthodox sciences. Mathematics was considered suitable as a pursuit for the educated scholar, and similarly physics, up to a point, especially as they contributed to the engineering works so characteristic of the centralized bureaucracy. The need of Chinese bureaucratic

31

society for great works of irrigation and water-conservation (flood protection and tax-grain transport, Pl. 2) meant that hydraulic engineering was regarded favourably among the traditional scholars and also that it helped in its turn to stabilize and support the form of society of which they were an essential part. Many students believe that the origin and development of feudal-bureaucratic society in China was at least partly dependent on the fact that from very early times the undertaking of great hydraulic engineering works tended to cut across the boundaries of the lands of individual feudal lords, and thus had the effect of concentrating all power in the centralized bureaucratic imperial State. By contrast with these forms of applied science, alchemy was distinctly unorthodox, the characteristic pursuit of disinterested Taoists and other recluses. Medicine was in this respect rather neutral; on the one hand the demands of traditional filial piety made it a respectable study for the scholars, while on the other its necessary association with pharmacy connected it with the Taoist alchemists and herbalists.

In the end I believe we shall find that the centralized feudal-bureaucratic style of social order was in the earlier stages favourable to the growth of applied science. Take the case of the seismograph which we have already mentioned more than once. It is paralleled by the existence of rain-gauges and even snow-gauges at a remarkably early time, and it is highly probable that the stimulus for such inventions came from the very reasonable desire of the centralized bureaucracy to be able to foresee coming events. Thus, for example, if a particular region was hit by a severe earthquake, it would be advisable to know this as soon as possible in order that help might be sent and reinforcements supplied to the local authorities in case of popular uprising. Similarly the rain-gauges set up on the edge of the Tibetan massif would have played a most useful part in determining the measures to be taken for the protection of hydraulic engineering works. Moreover Chinese society in the Middle Ages was able to mount much greater expeditions and pieces of organized scientific field work than was the case in any other medieval society. A good example of this would be the meridian arc surveyed early in the eighth century A.D. under the auspices of I-Hsing and the Astronomer-Royal, Nankung Yüeh. This geodetic survey covered a line no less than 2500 kms long reaching from Indo-China to the borders of Mongolia. About the same time an expedition was sent down to the East Indies for the purpose of surveying the constellations of the Southern Hemisphere within 20° of the south celestial pole. It is doubtful whether any other State in the world at that time could have engaged successfully in such activities.

From early times Chinese astronomy had benefited from State support but the semi-secrecy which it involved was to some extent a disadvantage. Chinese historians sometimes themselves realized this, for example in the dynastic history of the Chin dynasty (A.D. 265–420) there is a passage which says:

Astronomical instruments have been in use from very ancient days, handed down from one dynasty to another and closely guarded by official astronomers. Scholars therefore have had little opportunity to examine them, and this is the reason why unorthodox cosmological theories have been able to spread and flourish.'

However, one must not push this too far. It is quite clear that in the Sung period at any rate the study of astronomy was quite possible and even usual in scholarly families connected with the bureaucracy. Thus we know that in his earlier years Su Sung had model armillary spheres of small size in his home and so came gradually to understand astronomical principles. About a century later the great philosopher Chu Hsi also had an armillary sphere in his house and tried hard to reconstruct the water-power drive of Su Sung, although unsuccessfully. Besides, there were periods, for example in the eleventh century A.D., when mathematics and astronomy played quite a prominent part in the celebrated official examinations for the civil service.

INVENTION AND LABOUR POWER

We can now return to the question raised before, the relation of invention to labour power. Chinese labour conditions were no bar to a long series of 'labour-saving' inventions. Whether one thinks of the efficient trace-harness for horses (from the fourth century B.C. onwards) or the appearance of the still better collar-harness in the fifth century A.D., or of the simple wheelbarrow in the third century A.D. (though not in Europe till a thousand years later), one constantly finds that in spite of the seemingly inexhaustible masses of man-power in China, lugging and hauling was avoided whenever possible. How striking it is that in all Chinese history there is no parallel for the slave-manned oared war galley of the Mediterranean—land-locked though most of the Chinese waters were, sail was the characteristic motive power throughout the ages; and the arrival of great junks at Zanzibar or Kamchatka was only an extension of the techniques of the Yangtze and·the Tung-thing Lake. When the water-mill appeared in the first century A.D. for blowing metallurgical bellows the records concerning Tu Shih distinctly say that he considered it important as being both more humane and cheaper than man-power or animal-power, and it gives food for thought to find around A.D. 1300, four centuries before similar developments in Europe, water-power widely applied to textile machinery, especially silk-throwing and hemp-spinning.

All this is in considerable contrast to the position in Europe, where we do know of classical examples of refusal of innovation for fear of technological unemployment. I suppose that the best known example is the Roman one of the imperial refusal to make use of a machine for moving temple columns on the ground that it would put the porters out of work, and another case equally well established is that of the frame knitting machine in the seventeenth century A.D. It looks as if the Chinese example shows that shortage of labour may not in every culture be the sole stimulus for labour-saving inventions. Of course the problems here are very complicated and much further investigation is necessary.

PHILOSOPHICAL AND THEOLOGICAL FACTORS

It is clear that a full-dress comparison will eventually have to be made between the effects upon science and technology of the

Confucian–Taoist view of the world as compared with those of Christendom and Islam. Now the Confucian school which dominated the minds of the literati for more than 2000 years was primarily, as is generally known, this-worldly in character, occupied with a form of social ethics which purported to show the way whereby human beings could live in happiness and harmony together within society. The Confucians were concerned with human society and with what the West called natural law, that way of behaviour which it consorted with the actual nature of man that man should pursue. In Confucianism ethical behaviour partook of the nature of the holy, but it was not divine and had nothing to do with divinity since the conception of a creator God was unnecessary in the system. The Taoists, on the other hand, walked outside society; their Tao was the Order of Nature, not merely the order of human life, and it worked in a profoundly organic way in all its operations. Unfortunately, while the Taoists were extremely interested in Nature, they tended to distrust reason and logic, so that the workings of the Tao tended to remain somewhat inscrutable. Thus on the one hand interest was concentrated purely in human relations and social order, while on the other hand interest in Nature existed very strongly but tended to be mystical-experimental rather than rational-systematic.

No doubt one of the central features here is the contrast between conceptions of the laws of Nature in China and in the West. My colleagues and I have engaged in a rather thorough investigation of the concepts of laws of Nature in East Asian and West European culture. In Western civilization the ideas of natural law in the juristic sense and of the laws of Nature in the sense of the natural sciences can easily be shown to go back to a common root. Without doubt one of the oldest notions of Western civilization was that just as earthly imperial law-givers enacted codes of positive law to be obeyed by men, so also the celestial and supreme rational Creator Deity had laid down a series of laws which must be obeyed by minerals, crystals, plants, animals and the stars in their courses. There can be little doubt that this idea was intimately bound up with the development of modern science at the

Renaissance in the West. If it was absent elsewhere, could that not have been one of the reasons why modern science arose only in Europe; in other words, were medievally conceived laws of Nature in their naïve form necessary for the birth of modern science?

There can be little doubt that the conception of a celestial law-giver 'legislating' for non-human natural phenomena had its first origins among the Babylonians. The sun-god Marduk is pictured as the law-giver to the stars. This conception found continuation not so much among the pre-Socratics or the Peripatetics in Greece as with the Stoics, whose conception of universal law immanent in the world included non-human nature as much as man. In the Christian centuries the conception of the legislating Godhead was greatly increased by the stream of Hebrew influence. Throughout the Middle Ages the conception of divine legislation over non-human nature remained more or less a commonplace, but at the Renaissance the metaphor began to be taken very seriously indeed. The turning point occurred between Copernicus, who never used the expression 'law', and Kepler who did, though strangely enough not for any of his three great laws of planetary motion. It is curious to find that one of the very first applications of the expression 'law' to natural phenomena occurs not in astronomy or any of the biological sciences but in a geological-mineralogical context in one of the works of Agricola.

The Chinese world-view depended upon an entirely different line of thought. The harmonious co-operation of all beings arose, not from the orders of a superior authority external to themselves, but from the fact that they were all parts in a hierarchy of wholes forming a cosmic and organic pattern, and what they obeyed were the internal dictates of their own natures. Chinese conceptions of law did not develop the idea of laws of Nature for several different reasons. First, the Chinese early acquired a great distaste for precisely formulated abstract codified law from the abortive tyranny of the politicians belonging to the School of Legalists during the period of transition from feudalism to bureaucratism. Then when the bureaucratic system was finally set up, the old conceptions of natural law in the form of accepted customs and good *mores* proved more suitable than any others for Chinese society in its typical form, so that in fact the elements of natural law were much more important relatively in Chinese than in European society. But most of it was never put into formal legal terms, and since it was overwhelmingly human and ethical in content it was not easy to extend its sphere of influence to any form of non-human nature. Finally, and perhaps most important of all, the available ideas of a Supreme Being, though certainly present from the earliest times, became depersonalized so soon, and so severely lacked the idea of creativity, that they prevented the development of the conception of laws ordained from the beginning by a celestial law-giver for non-human nature. Hence the conclusion did not follow that other lesser rational beings could decipher or reformulate the laws of a great rational Super-Being if they used the methods of observation, experiment, hypothesis and mathematical reasoning. Of course this did not prevent the great development of science and technology in ancient and medieval China, many aspects of which we have already discussed, but it may have had a deep influence at the time of the Renaissance.

In the outlook of modern science there is, I presume, no residue of the notions of command and duty in the laws of Nature; they are now thought of as statistical regularities valid only in given times and places or for specified dimensions of size, descriptions not prescriptions. We dare not trespass here upon the great debate concerning subjectivity in formulations of scientific law, but the question does arise whether the recognition of statistical regularities and their mathematical expression could have been reached by any other road than that which Western science actually travelled. We might ask perhaps whether the state of mind in which an egg-laying cock could be prosecuted at law was necessary in a culture which should later on have the property of producing a Kepler.

THE LINGUISTIC FACTOR

Further study of the social relations of science and technology in China will of course also have to concern itself with the role of language. There is a commonly received idea that the ideographic language was a powerful inhibitory factor to the development of . modern science in China. We believe, however, that this influence is generally grossly overrated. It has proved possible in the course of our work to draw up large glossaries of definable technical terms used in ancient and medieval times for all kinds of things and ideas in science and its applications. Furthermore the Chinese language at the present day is found to be no impediment by the scientists of the contemporary culture. The National Academy at Peking (Academia Sinica) publishes today a wide range of scientific journals covering nearly all branches of research, and the language used today takes benefit from fifty years of work by the National Institute of Compilation and Translation which has defined technical terms for modern usage. We are strongly inclined to believe that if the social and economic factors in Chinese society had permitted or facilitated the rise of modern science there as well as in Europe, then already 300 years ago the language would have been made suitable for scientific expression.

At the same time also it is wise not to underestimate the capacities of the classical language. We do not recollect any instance where (after adequate consideration) we have been seriously in doubt as to what was intended by a classical or medieval Chinese author dealing with a scientific or technical subject, provided always that the text was not too corrupt, and that the description was sufficiently full. The general tendency was, of course, to make the descriptions too laconic. We often lack details because the literary scholars of later times abbreviated the records, not being themselves interested in scientific and technological matters. Similarly, the technical illustrations sometimes offer difficulties probably because the Confucian artists felt impatience at being asked to limn such inelegant and banausic objects. But where we do have sufficient details, as in the case of Su Sung's own description of his astronomical clock-tower in A.D. 1090, the *Hsin I Hsiang Fa Yao*, then it becomes possible to reconstruct down to the smallest detail what exactly was done.

Moreover the classical language is capable of a magnificently crystalline epigrammatic formulation which is not at all unsuitable for the best kind of philosophical thinking. Of this I may give an example from the works of Chu Hsi. In the twelfth century A.D., writing of his theory of organic development, a kind of emergent evolutionism or organic materialism recognizing a series of integrative levels, he said, 'Cognition or apprehension is the essential pattern of the mind's existence, but that there is something in the world which can do this is what we may call the spirituality inherent in matter.' To say all this took him only fourteen words: *So chio chê hsin chih li yeh; nêng chio chê, chhi chih ling yeh.* In other words, the mind's function is perfectly natural, something which matter has the potentiality of producing when it has formed itself into collocations with a sufficiently high degree of pattern and organization. The fact that Ma Chün could not talk like this, and explain his ideas to the supercilious scholars of the Chin court, only meant that he was neither a philosopher nor an orator, it did not mean that he could not explain to his own artisans exactly what he wanted made in the world of gear-wheels and link-work.

THE ROLE OF THE MERCHANTS

The last point which I must make here concerns the position of merchants in classical Chinese society. This brings us back to what I said before concerning the nature of feudal-bureaucratism. The institution of the mandarinate had the effect of creaming off the best brains of the nation for more than 2000 years into the civil service. Merchants might acquire great wealth yet they were never secure, they were subjected to sumptuary laws, and they could be mulcted of their wealth by inordinate taxation and every other kind of governmental interference. Moreover they never achieved a mystique of their own. In every age the sole ambition of the sons of even wealthy merchants was to get into the official bureaucracy; such was the prestige value of the world-outlook of the scholar-gentry that every channel of advancement led through it and every young man of whatever origin wanted to get into it. This situation prevailing, it was evidently impossible for the mercantile classes of

Chinese culture to acquire anything like the positions of power and influence in the State which they attained during the Renaissance in Europe. In other words, not to put too fine a point upon the matter, whoever would explain the failure of Chinese society to develop modern science had better begin by explaining the failure of Chinese society to develop mercantile and then industrial capitalism. Whatever the individual prepossessions of Western historians of science, all are necessitated to admit that from the fifteenth century A.D. onwards a complex of changes occurred; the Renaissance cannot be thought of without the Reformation, the Reformation cannot be thought of without the rise of modern science, and none of them can be thought of without the rise of capitalism, capitalist society and the decline and the disappearance of feudalism. We seem to be in the presence of a kind of organic whole, a packet of change, the analysis of which has hardly yet begun. In the end it will probably be found that all the schools, whether the Weberians, or the Marxists, or the believers in intellectual factors alone, will have their contribution to make.

The fact is that in the spontaneous autochthonous development of Chinese society no drastic change parallel to the Renaissance and the 'scientific revolution' in the West occurred at all. I often like to sketch the Chinese evolution as represented by a relatively slowly rising curve, noticeably running at a higher level, sometimes a much higher level, than European parallels between, say, the second and the fifteenth centuries A.D. But then after the scientific renaissance had begun in the West with the Galilean revolution, with what one might almost call the discovery of the basic technique of scientific discovery itself, then the curve of science and technology in Europe begins to rise in a violent, almost exponential, manner, overtaking the levels of the Asian societies and bringing about the state of affairs which we have seen during the past two or three hundred years. This violent disturbance of the balance is now beginning to right itself. No doubt in true historical thinking the 'ifs' so attractive to popular thought are out of place, but I would be prepared to say that if parallel social and economic changes had been possible in Chinese society then some form of modern science would have arisen there. If so, it would have been, I think organic rather than mechanical from the first, and it might well have gone a long way before receiving the great stimulus which a knowledge of Greek science and mathematics would no doubt have provided, and turning into something like the science which we know today. This is of course a question of the same character as 'if Caesar had not crossed the Rubicon', etc., and I only state it in this categorical form in order to convey some idea of the general conclusions which a prolonged study of Chinese scientific and technological contributions has induced in the minds of my colleagues and myself.

THE OLD WORLD ORIGINS OF THE NEW SCIENCE

Now I should like to return to the question raised at the beginning, and to go a little further into the distinction between modern science on the one-hand, and ancient and medieval science on the other. I shall thus have to deal somewhat more fully with certain points that have already been touched upon. As the contributions of the Asian civilizations are progressively uncovered by research, an opposing tendency seeks to preserve European uniqueness by exalting unduly the role of the Greeks and claiming that not only modern science, but science as such, was characteristic of Europe, and of Europe only, from the very beginning. For these thinkers the application of Euclidean deductive geometry to the explanation of planetary motion in the Ptolemaic system constituted already the marrow of science, which the Renaissance did no more than propagate. The counterpart of this is a determined effort to show that all scientific developments in non-European civilizations were really nothing but technology.

For example, our most learned medievalist has recently written:

'Impressive as are the technological achievements of ancient Babylonia, Assyria, and Egypt, of ancient China and India, as scholars have presented them to us they lack the essential elements of science, the generalized conceptions of scientific explanation and of mathematical proof. It seems to me that it was the Greeks who invented natural science as we know it, by their assumption of a permanent, uniform, abstract order and laws by means of which the regular changes observed in the world could be explained by deduction, and by their brilliant idea of the generalized use of scientific theory tailored according to the principles of non-contradiction and the empirical test. It is this essential Greek idea of scientific explanation, 'Euclidean' in logical form, that has introduced the main problems of scientific method and philosophy of science with which the Western scientific tradition has been concerned.'[1]

Again in a recent interesting and stimulating survey entitled *Science since Babylon* we read:

'What is the origin of the peculiarly scientific basis of our own high civilization? . . . Of all limited areas, by far the most highly developed, most recognizably modern, yet most continuous province of scientific learning, was mathematical astronomy. This is the mainstream that leads through the work of Galileo and Kepler, through the gravitation theory of Newton, directly to the labours of Einstein and all mathematical physicists past and present. In comparison, all other parts of modern science appear derivative or subsequent; either they drew their inspiration directly from the successful sufficiency of mathematical and logical explanation for astronomy, or they developed later, probably as a result of such inspiration in adjacent subjects. . . . Our civilization has produced not merely a high intellectual grasp of science but also a high scientific technology. By this is meant something distinct from the background noise of the low technology that all civilizations and societies have evolved as part of their daily life. The various crafts of the primitive industrial chemists, of the metallurgists, of the medical men, of the agriculturists—all these might become highly developed without presaging a scientific or industrial revolution such as we have experienced in the past three or four centuries.'[2]

Even the distinguished and enlightened author of *Science in History* writes (in correspondence):

'The chief weakness of Chinese science lay precisely in the field which most interested them, namely astronomy, because they never developed the Greek geometry, and perhaps even more important, the Greek geometrical way of seeing things which provided the Renaissance with its main intellectual weapon for the breakthrough. Instead they had only the extremely precise recurrence methods deriving from Babylonian astronomy, and these, on account of their

[1] A. C. Crombie: 'The significance of medieval discussions of scientific method for the scientific revolution', *Critical Problems in the History of Science*, ed. Marshall Clagett (Madison, Wisconsin, 1959), 79.

[2] D. J. de Solla Price: *Science Since Babylon* (New Haven, Connecticut, 1961).

exactitude, gave them a fictitious feeling of understanding astronomical phenomena.'[1]

Finally the author of a noted book, *The Edge of Objectivity*, says:

Albert Einstein once remarked that there is no difficulty in understanding why China or India did not create science. The problem is rather why Europe did, for science is a most arduous and unlikely undertaking. The answer lies in Greece. Ultimately science derives from the legacy of Greek philosophy. The Egyptians, it is true, developed surveying techniques and conducted certain surgical operations with notable finesse. The Babylonians disposed of numerical devices of great ingenuity for predicting the patterns of the planets. But no Oriental civilization graduated beyond technique or thaumaturgy to curiosity about things in general. Of all the triumphs of the speculative genius of Greece, the most unexpected, the most truly novel, was precisely its rational conception of the cosmos as an orderly whole working by laws discoverable in thought. ....'[2]

The statement of Einstein here referred to is contained in a now famous letter which he sent to J. E. Switzer of San Mateo, California, in 1953. It runs:

'Dear Sir,
The development of Western science has been based on two great achievements, the invention of the formal logical system (in Euclidean geometry) by the Greek philosophers, and the discovery of the possibility of finding out causal relationships by systematic experiment (at the Renaissance). In my opinion one need not be astonished that the Chinese sages did not make these steps. The astonishing thing is that these discoveries were made at all.

Sincerely yours,
Albert Einstein.'

It is very regrettable that this Shavian epistle with all its lightness of touch is now being pressed into service to belittle the scientific achievements of the non-European civilizations. Einstein himself would have been the first to admit that he knew almost nothing concrete about the development of the sciences in the Chinese, Sanskrit and Arabic cultures except that *modern* science did not develop in them, and his great reputation should not be

[1] J. D. Bernal.
[2] C. C. Gillispie: *The Edge of Objectivity; an Essay in the History of Scientific Ideas* (Princeton, N.J. 1960).

43

brought forward as a witness in this court. I find myself in complete disagreement with all these valuations and it is necessary to explain briefly why.

First, these definitions of mathematics are far too narrow. It would of course be impossible to deny that one of the most fundamental elements in Galileo's thinking was the geometrical study of kinematic problems. Again and again he praises the power of geometry as opposed to 'logic'. And geometry remained the primary tool for studying the problems of physical motion down to the early nineteenth century. But vast though the significance of deductive geometry was, its proofs never exhausted the power of the mathematical art. Although we speak of the Hindu–Arabic numerals, the Chinese were in fact the first, as early as the fourteenth century B.C., to be able to express any desired number, however large, with no more than nine signs. Chinese mathematics, developing the earlier Babylonian tradition, was always, as I have already said, overwhelmingly arithmetical and algebraical, generating such concepts and devices as those of decimal place-value, decimal fractions and decimal metrology, negative numbers, indeterminate analysis, the method of finite differences, and the solution of higher numerical equations. Very accurate values of π were early computed. The Han mathematicians anticipated Horner's method for obtaining the roots of higher powers. The triangle of binomial coefficients, as we have seen, was already considered old in the *Ssu Yuan Yü Chien* of A.D. 1303. Indeed in the thirteenth and fourteenth centuries A.D. the Chinese algebraists were in the forefront of advance as their Arabic counterparts had been in previous centuries, and so also the Indian mathematicians when they originated trigonometry (as we know it) nearly a thousand years earlier. To say that whatever algebra was needed by Vieta and by Newton they could easily have invented themselves may be uncritical genius-worship, but it is worse, it is unhistorical, for the influence of Asian ways of computation on European mathematicians of the later Middle Ages and the Renaissance is well established. And when the transmissions are examined the balance shows that between 250 B.C. and A.D. 1250, in spite of all China's isolations and inhibitions, a great deal more mathematical influence came out of that culture than went in.

Moreover the astronomical application of Euclidean geometry in the Ptolemaic system was not all pure gain. Apart from the fact (which some of these writers unaccountably seem to forget) that the resulting synthesis was in fact objectively wrong, it ushered the Western medieval world into the prison of the solid crystalline celestial spheres—a cosmology incomparably more naïve and *borné* than the infinite empty spaces of the Chinese *hun-thien* school or the relativistic Buddhist philosophers. It is in fact important to realize that Chinese thought on the world and its history was over and over again more boldly imaginative than that of Europe. The basic principles of Huttonian geology were stated by Shen Kua in the late eleventh century A.D., but this was only a counterpart of a Plutonian theme recurring since the fourth century A.D., that of the *sang thien* or mountains which had once been at the bottom of the sea. Indeed the idea of an evolutionary process, involving social as well as biological change, was commonly entertained by Chinese philosophers and scientifically interested scholars, even though sometimes thought of in terms of a succession of world renewals following the catastrophes and dissolutions assumed in the recurrent *mahākalpas* of Indian speculation. One can see a striking echo of this open-mindedness in the calculations made by I-Hsing about A.D. 724 concerning the date of the last general conjunction. He made it come out to 96,961,740 years before—rather a different scale from '4004 B.C. at six o'clock in the evening'.

Thirdly, the implied definitions of science are also much too narrow. It is true that mechanics was the pioneer among the modern sciences, the 'mechanistic' paradigm which all the other sciences sought to imitate, and emphasis on Greek deductive geometry as its base is so far justifiable. But that is not the same thing as saying that geometrical kinematics is all that science is. Modern science itself has not remained within these Cartesian bounds, for field theory in physics and organic conceptions in biology have deeply modified the earlier mechanistic world-picture. Here knowledge

45

of magnetic phenomena was all-important, and this was a typically Chinese gift to Europe. Although we do not know the way-stations through which it came, its priority of time is such as to place the burden of proof on those who would wish to believe in an independent discovery. The fact is that science has many aspects other than geometrical theorizing. To begin with, it is nonsense to say that the assumption of a permanent, uniform, abstract order and laws by means of which the regular changes in the world could be explained, was a purely Greek invention. The order of Nature was for the ancient Chinese the *Tao*, and as a *chhang Tao* it was an 'unvarying Way'. 'Every natural phenomenon,' says the fourth-century B.C. *Chi Ni Tzu* book, 'the product of Yin and Yang, has its fixed compositions and motions with regard to other things in the network of Nature's relationships.' 'Look at things,' wrote Shao Yung in the eleventh century A.D., 'from the point of view of things, and you will see their true nature; look at things from your own point of view, and you will see only your own feelings; for nature is neutral and clear, while feelings are prejudiced and dark.' The organic pattern in Nature was for the medieval Chinese the *Li*, and it was mirrored in every subordinate whole as one or another *wu li* of particular things and processes. Since the thought of the Chinese was in all ages profoundly organic and impersonal they did not envisage laws of a celestial lawgiver—but nor did the Greeks, for it is easily possible to show that the full conception of Laws of Nature attained definitive status only at the Renaissance.

What the Chinese did do was to classify natural phenomena, to develop scientific instruments of great refinement for their respective ages, to observe and record with a persistence hardly paralleled elsewhere, and if they failed (like all medieval men, Europeans included) to apply hypotheses of modern type, they experimented century after century obtaining results which they could repeat at will. When one recites this list of the forms of scientific activity it becomes difficult to see how anyone could deny them their status as essential components of fully developed world science, biological and chemical as well as astronomical and physical, if it was not in the interest of some instinctive *parti pris*.

Elaborating, *kho hsüeh*, the traditional and current Chinese term for science, means 'classification knowledge'. The first star catalogues, probably pre-Hipparchan, open its story in China. It is then exemplified in the long series of rational pharmacopœias which begins with the second-century B.C. *Shen Nung Pên Tshao*. It helped to lay the basis of our knowledge of chemical affinity in the theories of polarities (*i*) and categories (*lei*) found in treatises such as the fifth-century A.D. *Tshan Thung Chhi Wu Hsiang Lei Pi Yao*. If systematic classifications of parhelic phenomena in the heavens (*Chin Shu*), and of the diseases of men and animals on earth (*Chu Shih Ping Yuan*), were worked out a full thousand years before Scheiner and Sydenham, this was only the expression of the firm hold which the Chinese had on this basic form of scientific activity. Perhaps the view of science which I am criticizing rests partially on too great a preoccupation with astronomy. and too little with biology, mineralogy and chemistry.

Then as to apparatus. That the Hellenistic Greeks were capable of producing highly complicated scientific instruments is shown by the anti-Kythera computing machine, but this is a very rare, indeed a unique example. It would be fairer to admit that throughout the first fifteen centuries of our era Chinese instrument-making was generally ahead, and (as in such instances as the seismograph and the mechanical clock) often much ahead, of anything that Europe could show. Actually the invention of clockwork was directly connected with the very absence of planetary models in Chinese thinking, for while on ecliptic co-ordinates no real body ever moves, declination circles are tracks of true motion, and the equatorial-polar system was a direct invitation to construct planetaria mechanically rotated. So, too, modern positional astronomy employs not the ecliptic co-ordinates of the Greeks but the equatorial ones of the Chinese. Nor need we confine ourselves, to the astronomical sciences here, for a wealth of advanced techniques is to be found in those alchemical treatises of which the *Tao Tsang* is full.

Surely, again, observation, accurate and untiring, is one of the foundation-stones of science. What records from an antique culture are of vital interest to radio-astronomers today? Nothing from Greece, only the nova, comet and meteor lists of China's star-clerks. They it was who first established (by the seventh century A.D. at least) the constant rule (*chhang tsê*) that the tails of comets point away from the sun. Renaissance astronomers who quarrelled so much among themselves about the priority of the study of sun-spots might have been somewhat abashed if they had known that these had been observed since the first century B.C. in China, and not only observed but recorded in documents reliably handed down. When Kepler penned his New Year letter on the hexagonal form of snowflake crystals in A.D. 1611 he did not know that his contemporary Hsieh Tsai-Hang was puzzling over just the same thing, not, however, as a new idea but as a fact which had been known and discussed since the original discovery reported by Han Ying in the second century B.C. When we look for the original root of the cloud-seeding process in the comparison of snow-flake crystals with those of various salts and minerals, we find it not in the eighteenth-century A.D. experiments of Wilcke but in the acute observation of Chu Hsi in the twelfth century A.D. Thus it will surely be apparent that if God could geometrize so could the Tao, and the Europeans were not the only men who could trace its operations in forms both living and non-living. Finally if an example is needed from the biological sciences, let us remember the brilliant empirical discovery of deficiency diseases clearly stated by the physician Hu Ssu-Hui in the fourteenth century A.D.

Degree of accuracy in observation is also relevant. Indeed it is a vital feature, for it springs from that preoccupation with quantitative measurement which is one of the most essential hallmarks of true science. The old astronomical lists gave stellar positions in measured degrees, of course, the hydraulic engineers were recording precisely the silt-content of rivers in the first century B.C., and the pharmacists early developed their systems of dosages, but another example, less known, is more striking. Of the dial-and-pointer readings which make up so much of modern science, a search throughout the medieval world between the eighth and the fourteenth centuries A.D. would reveal instruments capable of giving them only in China. I refer to the needles of the magnetic compasses used first by geomancers, then (at least a century before Europe) by the sea-captains. Now it is a remarkable fact (as we have seen) that the Chinese were worrying about the cause of magnetic declination for a considerable time before Europeans knew even of magnetic direct: vity. Indeed the geomantic compass in its final form embodies two additional rings of points, one staggered $7\frac{1}{2}°$ east and the other $7\frac{1}{2}°$ west—these represent the remains of observations of declination, eastwards before about A.D. 1000 and westwards thereafter. We have reason to believe that this disturbing discovery was first made some time in the ninth or tenth centuries A.D., and it could never have been made if the observers had not been marking with extraordinary accuracy—and honesty—the 'true path' of the needle. It is even legitimate to compare this feat in principle with the discovery of the inert or noble gases so long afterwards by Rayleigh and Ramsay, residual bubbles which others had put down to experimental error or simply neglected. The honesty deserves emphasis also, for it was not shown so clearly when Europeans came up against the same phenomenon four or five centuries later. Or one might say that they had a greater tolerance of error, being content with 'there or thereabouts'. The history of magnetic declination in the West has been obscured by the fact that the compass-makers 'fiddled' the instrument by fixing a card askew to make it read right, and little or nothing was written about the matter till the sixteenth century A.D. Similarly, Robert Norman used to 'fiddle' his compasses to make the needles lie horizontally, until one day he lost his temper and really looked into the trouble, so re-discovering 'dip' or inclination.

Perhaps the greatest objection to the attempt of the Hellenizers to save European superiority is the fact that the Greeks were not real!y experimenters. Controlled experimentation is surely the greatest methodological discovery of the scientific revolution of the Renaissance, and it has never been convincingly shown that any earlier group of Westerners fully understood it. I do not propose to claim this honour for the medieval Chinese either, but they came just as near it theoretically, and in practice often went beyond European achievements. Although the ceramics technologists of China undoubtedly paid great attention to their temperatures and to the oxidizing-reducing atmospheres of their kilns, I shall not return to this here, for the Hellenizers would no doubt include the immortal products of the Sung potters in that "background noise of low technology' which was all that non-European cultures could attain. I prefer, then, to take other examples: Tu Wan's labelling of fossil brachiopods ('stone-swallows') to demonstrate that if they ever flew through the air it was only to drop down by process of weathering, or the long succession of pharmaceutical experiments on animals carried out by the alchemists from Ko Hung to Chhen Chih-Hsü, or the many trials made by the acoustics experts on the resonance phenomena of bells and strings, or the systematic strength-of-material tests which internal evidence shows must have been undertaken before the long beam bridges across the Fukienese estuaries could have been constructed. Is it possible to believe that apparatus so complex as that of the water-wheel linkwork escapement clocks, or indeed much of the textile machinery, could ever have been devised without long periods of workshop experimentation? The fact that written records of it have not come down to us is only what we should expect in a medieval literary culture. The fact that none of it was carried out on isolated and simplified objects, such as balls rolling down inclined planes, is again only what was characteristic of pre-Renaissance practice everywhere.

I do not say that the Greek *praeparatio evangelica* was not an essential part of the background of modern science. What I do want to say is that modern exact and natural science is something much greater and wider than Euclidean geometry and Ptolemaic mathematical astronomy; more rivers than those have emptied into its sea. For anyone who is a mathematician and a physicist, perhaps a Cartesian, this may not be welcome; but I myself am professionally a biologist and a chemist, more than half a Baconian, and I therefore do *not* think that what constituted the spearhead of the Galilean break-through constitutes the whole of science. What happened to crystallize the mathematization of experimental hypotheses when the social conditions were favourable does not exhaust the essence. If mechanics was the primary science, it was *primus inter pares*. If physics celestial and terrestrial has the battle-honours of the Renaissance, it is not to be confused with the whole army of science, which has many brave regiments besides.

'The spearhead, but not the whole, of science.' In pondering over a better way of representing the situation, it occurred to me that we ought perhaps to make a clearer distinction between factors which were concerned in the direct historical genesis of modern science, and factors which fell into place later after the Galilean break-through. We shall also have to distinguish more clearly between science and technology. Suppose we erect a classi-fication of four pigeon-holes, science vertically on the left and technology vertically on the right, and let the upper boxes represent direct historical genesis while the lower ones represent subsequent reinforce-ment. Then taking the upper left-hand compart-ment first, the contribution of the Greeks will have the greatest share, for Euclidean deductive geometry and Ptolemaic astronomy, with all that they imply, were undoubt-edly the largest factor in the birth of the 'new, or experimental, science'—in so far as any antecedents played a part at all, for we must not undervalue its basic originality. In spite of Ptolemy and Archimedes, the occidental ancients did not, as a whole, experi-ment. But Asian contributions will not be absent from this compartment, for not only must we leave a place for algebra and the basic numerational and computational techniques, we must not forget the significance of magnetism, and knowledge of this realm of phenomena had been built up exclusively in the Chinese culture-area, which thus powerfully influenced Europe through Gilbert and Kepler. Here one remembers also the adoption of the Chinese equatorial co-ordinates by Tycho. But the Greeks pre-dominate. In the upper right-hand compartment the situation is entirely different, for in technology Asian influences in and before the Renaissance (especially Chinese) were legion—I need mention only the efficient horse-harnesses, the technology of iron and steel, the inventions of gunpowder and paper, the escapement of the mechanical clock, and basic engineering devices such as the driving-belt, the chain-drive, and the standard method of converting rotary to rectilinear motion, together with nautical techniques such as the leeboard and the stern-post rudder. Alexandria also ran.

The lower compartments will now be available to take achievements of the Asian cultures which, though not genetically connected with the first rise of modern science yet deserve all praise; they may or may not be directly genetically related to their corresponding developments in post-Renaissance modern science. A case of direct influence could be found in the Chinese doctrine of infinite empty space instead of solid crystalline celestial spheres, but it did not operate until after Galileo's time. Cases of later incorporation would be the development of undulatory theory in eighteenth-century A.D. physics, which immensely elaborated characteristically Chinese ideas without directly building on them, or the use of ancient and medieval Chinese records by radio-astronomers. So also, if atomism, not mathematics, proved to be the soul of chemistry, which found itself so much later than physics, this elaborated Indian and Arabic ideas of great subtlety without knowingly basing itself thereon. A good case of the absence of any influence would be the seismograph as used in China from the second to the seventh centuries A.D.; though an outstanding achievement, it was almost certainly unknown to any of the scientific men who developed seismographs in post-Renaissance Europe. Chinese biological and pathological classification. systems occupy the same position; they were clearly unknown to Linnaeus and Sydenham, but none the less worthy of study, for only by drawing up the balance-sheet in full shall we ever ascertain what each civilization contributed to human advancement. It is not legitimate to require of every scientific or technological activity that it should have contributed to the advancement of the

European culture-area. What happened in other civilizations is entirely worth studying for its own sake. Must the history of science be written solely in terms of one continuous thread of linked influences? Is there not an ideal history of human thought and knowledge of nature, in which every effort can find its place, irrespective of what influences it received or handed on? Modern universal science and the history and philosophy of universal science will embrace all in the end.

It only remains to consider the contents of the right-hand lower compartment. Here we have to think of technical inventions which only became incorporated, whether or not by re-invention, into the corpus of modern technology after the Renaissance period. A case in point might be the paddle-wheel boat, but it is uncertain, for we do not know whether the first European successes were based on a Byzantine idea never executed, or on a vast fund of practical Chinese achievement during the preceding millennium. A better case would be the differential gear, for though present in the south-pointing carriage of ancient China, it must almost certainly have arisen again independently in Europe. So also the Chinese methods of steel-making by the co-fusion process and by the direct oxygenation of cast iron, though of great seniority to the siderurgy of Europe, were not able to exert any influence upon it, if indeed they did, which is still uncertain, until long after the Renaissance. Similarly it might be unwise to connect too closely the crucible steel of Huntsman with that of the age-old Indian wootz process.

In all this I have tried to offer an *opinio conciliatrix* in friendly fashion to those who may have been shocked by the objective attitude which I always seek to adopt in weighing European claims. If we think out the matter as I suggest, we may feel greater need for recognizing several kinds of values; the value of that which helped directly to effect the Galilean break-through, the value of that which became incorporated in modern science later on, and last but not least, the value of that residue which yet renders other civilizations no less worthy of study and admiration than Europe.

The erroneous perspective which I am criticizing can be seen particularly well in the use of the possessive plural personal pronoun. Some Western historians of science constantly speak of 'our modern culture' and 'our high civilization' (I italicize). *The Edge of Objectivity* reveals even more clearly the mood in which they approach the comparative study of men's efforts to understand and control the natural world.

'Anxious though our moments are, today is not the final test of wisdom among statesmen or virtue among peoples. The hard trial will begin when the instruments of power created by the West come fully into the hands of men not of the West, formed in cultures and religions which leave them quite devoid of the Western sense of some ultimate responsibility to man in history. The secular legacy of Christianity still restrains our world in some slight measure, however self-righteous it may have become on one side and however vestigial on the other. Men of other traditions can and do appropriate *our* science and technology, but not our history or values. And what will the day hold when China wields the bomb? And Egypt? Will Aurora light a rosy-fingered dawn out of the East? Or Nemesis?'

This is certainly very near the edge. It would induce in the reader a lamentable and unworthy attitude of mind in which fear would jostle its counterpart, possessiveness. Surely it would be better to admit that men of the Asian cultures also helped to lay the foundations of mathematics and all the sciences in their medieval forms, and hence to set the stage for the decisive break-through which came about in the favourable social and economic milieu of the Renaissance. Surely it would be better to give more attention to the history and values of these non-European civilizations, in actual fact no less exalted and inspiring than our own. Then let us give up that intellectual pride which boasts that 'we are the people, and wisdom was born with us'. Let us take pride enough in the undeniable historical fact that *modern* science was born in Europe and only in Europe, but let us not claim thereby a perpetual patent thereon. For what was born in the time of Galileo was a universal palladium, the salutary enlightenment of all men without distinction of race, colour, faith or homeland, wherein all can qualify and all participate. Modern universal science, yes; Western science, no!

2

SCIENCE AND CHINA'S INFLUENCE
ON THE WORLD

First published in *The Legacy of China*, Oxford, 1964.

I

In all civilizations it has been customary for those who are departing from this life to designate others as the heirs and inheritors of their worldly goods, and if such an action has the approbation of law we are accustomed to call what is thus transmitted a legacy. For the earlier volumes of the 'Legacy' series the term was appropriate enough, because the civilizations and the languages of classical Greece and imperial Rome have indeed long been dead in the historical sense, but difficulties naturally arise when the expression is continued for a civilization which is still living. If it is true that Sanskrit itself is a dead language, Indian culture as a whole is very much alive, and the same is true of the culture of China. In fact Chinese civilization has never been living more vigorously than it is today.

Who was supposed to be the 'legatee' in the original conception of the 'Legacy' series? Was not the unspoken conception that it was 'modern Western' civilization, rapidly spreading over the world, and destined to supersede all cultures not based on European Christendom? A 'Legacy of Europe' was not thought worth writing, for Europe alone was the repository of immortal truth.[1] It

---

[1] It is still not at all difficult to find expressions of this attitude towards the European inheritance by Western scholars. For example Costa Brochado has written: 'Truly the motive power of the forces which are working against us in

seems very doubtful now whether this implicit assumption has any validity. Modern science has indeed created a universal and international culture, that of the airman, the engineer, and the biologist, but although modern science originated in Europe and in Europe only, it was built upon a foundation of medieval science and technology much of which was non-European. Thus we must define our terms. The 'testator' here is a civilization with a longer continuous living tradition than any other, with the possible exception of Israel, and one which is in no danger of decay. The 'legatee' is the international world of which every country now inevitably forms a part; not simply a Europe deigning to adopt a few exotic elements from peripheral peoples. Each people enters the modern world with its own offering of thought, discovery, and invention, some richer perhaps than others, but each able and willing to participate in the universal discourse of applied mathematics while yet most faithful to its own inheritance of language and philosophy, an inheritance from which all others have much to learn. Indeed, the metaphor of inheritance is unsatisfactory for our purposes, for the inheriting process has been a series of mutual transmissions going on for more than twenty centuries. One would rather prefer the image of the great rivers of past science and technology flowing into the ocean of modern natural knowledge, so that all peoples have been testators and all are now inheritors, each in their several ways.

Asia and Africa today is a millenary hatred borne by Asiatic orientalism against the last torches of occidental civilization which still burn in those regions. What it cannot abide is the strong and redeeming light of a culture which proclaims the civil and moral liberty of man against that caste tyranny which is at the base of all oriental philosophies. This is the explanation of how it has been possible for the wonderful technology of Western civilization to be completely assimilated, from the frontiers of Asia to the confines of Japan, without any modification of the philosophic-religious conceptions of human life among those peoples. It looks as if China, Russia, India, and Japan, to name only the greater countries, have profited exclusively by the experimental sciences of our civilization in order to arm and equip themselves to destroy in the end all that is profound and essential in it, its spirit and its morality.' (*Henri le Navigateur*, Lisbon, 1960, p. 34.) Let it not be thought that this European chauvinism is confined to the Portuguese or other Europeans; a very similar statement can be found in C. C. Gillispie, *The Edge of Objectivity: an Essay in the History of Scientific Ideas* (Princeton, 1960), p. 8.

Secondly, the history of science and technology is not to be bounded by Europe and what Europe received of these transmissions. During the first fourteen centuries of the Christian era, as this contribution will have no difficulty in showing,[1] China transmitted to Europe a veritable abundance of discoveries and inventions which were often received by the West with no clear idea of where they had originated. The technical inventions of course travelled faster and further than the scientific thought. But besides all this there were important influences upon nascent modern science during the Renaissance period, and these continued on throughout the eighteenth century. By that time we reach the beginning of the modern period, when science has become a world-wide enterprise in which China is participating along with all other cultures. If we were to interpret the term 'legacy' in what was probably its original sense, we should restrict our attention to those factors which were concerned in the direct historical genesis of modern science, excluding factors which fell into place after the Galilean break-through of the early seventeenth century. But if we take the term in the wider sense outlined above, we shall be interested in what China contributed to the world at all periods. To explain what this means we must not only consider pre-Renaissance and post-Renaissance transmissions separately, but also distinguish between science and technology.

There can be no doubt that in the opening phases of modern science, when mechanics, dynamics, and celestial and terrestrial physics came into being in their modern form, the Greek contribution had the greatest share.[2] Euclidean deductive geometry and Ptolemaic planetary astronomy, with all that they imply, were certainly the main factors in the birth of the 'new, or

[1] It will of course be impossible to give detailed chapter and verse for any of the statements made in this paper. Abundance of references to both Chinese and Western literatures, together with Chinese characters, will however be found in *Science and Civilisation in China* (7 vols. in 12 parts, Cambridge, 1954–), by J. Needham, with the collaboration of Wang Ling (Wang Ching-Ning), Lu Gwei-Djen, Ho Ping-Yü, Kenneth Robinson, Tshao Thien-Chhin, and others, I am glad to record here my great indebtedness to my Chinese friends and co-workers, without whom the work would have been impossible.

[2] For a discussion of this point see the preceding paper.

experimental, science'—in so far as any antecedents played a part at all, for we must not underrate its basic originality. In spite of Ptolemy and Archimedes, the occidental ancients did not, as a whole, experiment. But Asian contributions were by no means absent from the decisive break-through, for apart from algebra and the basic numerational and computational techniques (for example the Indian numerals, the Indo-Chinese zero, and Chinese decimal place-value, the most ancient form of the method), China provided all the basic knowledge of magnetical phenomena. This field of study (to which we shall return presently, p. 71) was radically different from those which Greek physics had cultivated, and its effect upon the initial stages of modern science, mediated through Gilbert and Kepler, was of vital importance. There were also significant influences from China in practical astronomy, such as the adoption of the Chinese celestial co-ordinates by Tycho Brahe.

In technological influences before and during the Renaissance China occupies a quite dominating position. In the body of this contribution we shall mention among other things the efficient equine harness, the technology of iron and steel, the inventions of gunpowder and paper, the mechanical clock, and basic engineering devices such as the driving-belt, the chain-drive, and the standard method of converting rotary to rectilinear motion, together with segmental arch bridges and nautical techniques such as the stern-post rudder. The world owes far more to the relatively silent craftsmen of ancient and medieval China than to the Alexandrian mechanics, articulate theoreticians though they were.

We have next to think of those achievements of Asian and Chinese science which, though not genetically connected with the first rise of modern science, yet deserve close attention. They may or may not be directly related genetically to their corresponding developments in post-Renaissance modern science. Perhaps the most outstanding Chinese discovery which was so related, even though it influenced the West relatively late (the end of the eighteenth and the beginning of the nineteenth centuries), was that of the first successful immunization technique. Variolation, the forerunner of Jennerian vaccination, had been in use in China certainly since the beginning of the sixteenth century and, if tradition is right, since the eleventh; it consisted in the inoculation of a minute amount of the contents of the smallpox pustule itself into the nostril of the patient to be immunized, and Chinese physicians had gradually worked out methods of attenuating the virus so as to give greater safety. The origins of the whole science of immunology lie in a practice based on medieval Chinese medical thought. A case of direct theoretical influence which springs to mind concerns cosmology—the old Chinese doctrine of infinite empty space as opposed to the solid crystalline celestial spheres of medieval Europe, but again it did not exert its full effect towards their dissolution until after Galileo's time. Examples of later incorporation would be the development of undulatory theory in eighteenth-century physics, which immensely elaborated characteristically Chinese ideas without knowing anything of them; or the use of ancient and medieval Chinese records of novae and supernovae by modern radio-astronomers. A good case of the probable absence of any stimulus would be the seismograph as used in China from the second to the seventh centuries A.D.; though an outstanding achievement and a permanent legacy to the history of geology, it was almost certainly unknown to any of the scientific men who developed seismographs again in post-Renaissance Europe. Chinese biological and pathological classification systems occupy the same position; they were clearly unknown to Linnaeus and Sydenham, but none the less worthy of study, for only by drawing up the balance-sheet in full shall we ever ascertain what each civilization has contributed to human advancement. Similarly, it is now becoming clear that medieval Chinese anatomy was far more advanced than has generally been thought, for judgments have been based by Western anatomists only on the few remaining block-print illustrations, since they were unable to read the texts themselves and to pursue the complex and elaborate nomenclature. But it exerted no influence on the revival and development of anatomy in Renaissance Europe. Nor did the outstandingly good iconographic tradition of the pharmaceutical compendia of the *Pên Tshao* genre, centuries ahead of the West in accurate botanical illustration, which has gained appreciation only in our own time.

Lastly we have to think of technical inventions which only became incorporated, whether or not by re-invention, into the corpus of modern technology after the Renaissance period. A case in point might be the paddle-wheel boat, but it is uncertain, for we do not know whether the first European successes were based on a Byzantine idea never executed, or on a vast fund of practical Chinese achievement during the preceding millennium, or on neither. A clearer example is the iron-chain suspension bridge, for while the first European description came towards the end of the sixteenth century, the first realization occurred only in the eighteenth, and in knowledge of the Chinese antecedents, going back, as we now know, for more than a thousand years previously. Independent invention occurred, no doubt, with the differential gear, for though this was present in the south-pointing carriages of ancient China, their construction has been revealed only by modern historical research and could hardly have inspired the later mechanics of the West who fitted up again this important form of enmeshing wheel-work. So also the Chinese methods of steel-making by the co-fusion process and by the direct oxygenation of cast iron, though of great seniority to the siderurgy of Europe, were not able to exert any influence upon it, if indeed they did, which is still uncertain, until long after the Renaissance. At the same time one must always refrain from being too positive about the absence of influence. In human intercourse there have been innumerable capillary channels which we cannot see, and especially for earlier times we should never be tempted to dogmatism in the denial of transmissions. Sometimes one wonders whether humanity ever forgets anything. The sailing-carriage of early seventeenth century Europe was consciously modelled on supposed Chinese prototypes which had in fact been rather different, but it is possible that they in their turn derived from the model boats with sails outspread which, supported upon low wooden wagons, conveyed the coffins of ancient Egyptian gods or kings across the deserts to their tombs. Broadly speaking, experience shows that the further one goes back in history the more unlikely independent invention was; we cannot infer it from the conditions of modern science today, where it frequently occurs.

Thus in relation to the 'legacy' of China we have to think of three different values. There is the value of that which helped directly to effect the Galilean break-through, the value of that which became incorporated in modern science later on, and last but not least the value of that which had no traceable influence and yet renders Chinese science and technology no less worthy of study and admiration than that of Europe. Everything depends on the definition of the legatee—Europe alone, or modern universal science, or the whole of mankind. I would urge that it is not in fact legitimate to require of every scientific or technological activity that it should have contributed to the advancement of the European culture-area. Nor need it even be shown to have constituted building material for modern universal science. The history of science is not to be written solely in terms of one continuous thread of linked influences. Is there not an oecumenical history of human thought and knowledge of nature, in which every effort can find its place, irrespective of what influences it received or handed on? Is not the history and philosophy of universal science the only true legatee of all human endeavour?

II

So much for indebtedness and its various meanings. Misunderstandings have been corrected, but nothing intrinsically new has emerged. Yet in this paper I do want to make an important, if paradoxical, point which I do not remember seeing fully brought out anywhere hitherto. The proper title for this paper ought to be 'The Ten (or the Twenty or Thirty) Discoveries (or Inventions) that Shook the World'. That Chinese discoveries and inventions there were, we have long known; that they were transmitted one after the other to Europe, we can demonstrate or show to be extremely likely; but the extraordinary paradox arises that while many, even most, of them had earth-shaking effects upon occidental society, Chinese society had a strange capacity for absorbing them and remaining relatively unmoved. I shall return to this at the conclusion, having systematically pointed out the social effects of the Chinese novelties as we speak of them; perhaps to offer some tentative explanation of the outstanding contrast. Here I wish only to strike the real keynote of this paper.

One common misconception which it is desirable to get out of the way before going further is that Chinese achievements were invariably technical rather than scientific. It is true, as has already been said, that ancient and medieval Chinese science was hemmed in within the boundaries of the ideographic language and penetrated little outside it. But because the practical inventions were the only things that the Indian, Arabic, or Western cultures were generally capable of taking over from the Chinese culture-area, this does not mean that the Chinese themselves had always been mere 'sooty empiricks'. On the contrary, there was a large body of naturalistic theory in ancient and medieval China, there was systematic recorded experimentation, and there was a great deal of measurement often quite surprising in its accuracy. Of course, the theories of the Chinese remained to the end of their autochthonous period characteristically medieval in type, for the Renaissance, with its mathematization of hypotheses, did not happen among them.

The point can be illustrated perhaps by a quotation which in any case would only inexcusably be omitted from this chapter.

'It is well to observe [said Lord Verulam] the force and virtue and consequences of discoveries. These are to be seen nowhere more conspicuously than in those three which were unknown to the ancients, and of which the origin, though recent, is obscure and inglorious; namely, printing, gunpowder, and the magnet. For these three have changed the whole face and state of things throughout the world, the first in literature, the second in warfare, the third in navigation; whence have followed innumerable changes; insomuch that no empire, no sect, no star, seems to have exerted greater power and influence in human affairs than these mechanical discoveries.'[1]

[1] Francis Bacon, *Novun Organum*, bk. 1, aphorism 129.

Subsequent scholars, who might have been expected to know better, were content that the origin of the discoveries should remain obscure and inglorious. J. B. Bury, for instance, when describing the Renaissance controversies between the supporters of the 'Ancients' and those of the 'Moderns', shows that the latter were generally considered to have had the best of it, precisely because of the three great inventions which Bacon described. Yet nowhere in his book is there even a footnote pointing out that none of the three was of European origin.[1] In what follows we shall show not only where their origin was, but how it arose from ancient scientific theorizing.

Bury's book was written nearly fifty years ago, but the same attitude of 'invincible ignorance' about non-European contributions persists today as strongly as ever. One cannot help noticing it in a recent work, *The Inspiration of Science*, by Sir George Thomson.[2] After emphasizing the twin Greek successes of geometry and planetary astronomy, he goes on to say:

'But with things on earth they were less successful. They knew that amber when rubbed attracted chaff, and that a stone from a place in Asia Minor called Magnesia attracted iron, and they had observed that a pole sticking out of the water seemed bent; but they made no real progress with the corresponding sciences. It is sometimes said that this failure was due to an unwillingness to experiment. No doubt up to a point this is true, but I think that there is something more. . . . [The Greeks] failed to realize the importance of these apparently trivial occurrences. The heavens were impressive and grand, perhaps the abode of gods or even of something greater than the gods. Little bits of chaff and shreds of iron were amusing but hardly of the first importance. This is a very natural attitude.

Yet the greatest discovery in method which science has made is that the apparently trivial, the merely curious, may be clues to an understanding of the deepest principles of Nature. One can hardly blame the Greeks. Even with Newton behind him, Swift could be witty at the expense of the Royal Society in his account of the 'projectors' of Laputa, with their studies of cucumbers as a source of sunlight—and Swift, though an unpleasant creature, was no fool. Just how the discovery came about is not clear. It is the great thing that marks off our age from others, and may well have had several independent causes. Among

[1] *The Idea of Progress* (London, 1920), pp. 40 ff., 45, 54, 62, 78 ff., 138.
[2] Oxford, 1962.

these, probably, was the importance of magnetism for navigation, and of optics for spectacles. Gunnery perhaps added a little, and made Galileo's mechanics sound rather less improbable. But a greater cause was the excitement that came from the discovery of a way round Africa to India and then of the New World. In an age in which the wildest projects of geographical discovery had proved successful it was natural to try others of a different kind, to open the mind and ask more searching questions on matters nearer at hand. The first discovery must always be that there are things worth discovering. So the apparent trivialities of the stone from Magnesia and of amber grew in importance, and since the time of Maxwell it has been clear to the discerning that the ideas behind them are as fundamental as any in the world, not even excluding that of matter.'

Much of this is well said, and worth saying; but some of it is surely an example of what Claude Roy has called 'the iron curtain of false enigmas'. Not only is the Chinese origin of magnetical science and of explosives chemistry quietly ignored, but the beginning of curiosity about apparently trivial natural occurrences is made into a mystery. It may possibly be that the Greeks lacked this[1]—if so, they were already infected with that false sense of values which led Thomas Aquinas to say that 'a little knowledge about the highest things is better than the most abundant knowledge about things low and small'.[2] If the secrets of the Magnesian stone were first revealed in China this was perhaps not only due to the organic materialism of her cosmology, but also because all Chinese philosophical tradition lay behind Chhêng Ming-Tao's eleventh-century criticism of the Buddhists: 'When they strive only to "understand the high" without "studying the low", how can their understanding of the high be right?'[3]

III

Let us now return to the inventions listed by Bacon. Since everything cannot be discussed, we shall leave the epic of printing on

[1] But I should be surprised if it were not possible to find statements in the Hippocratic corpus and in Galen about the great importance of very small pathological symptoms or anatomical structures.

[2] *Summa Theologiae*, Ia, i. 5 ad 1.

[3] *Honan Chhêng shih I Shu*, ch. 13, p. 1b. The quoted phrases are from the *Analects* XIV. 37, where Confucius says that the study of the lowly elucidates the supernal (*hsia hsüeh, erh shang ta*). What a contrast to the theme of 'suspiciendo despicio'!

one side,[1] and stay only upon the discovery of chemical explosive force and that of magnetic polarity. It is hard to overrate either of them, and both developed from Taoist (originally Shamanist) magic, guided into practical reproducibility by the theories of Chinese natural philosophy working within alchemy on the one hand and geomancy on the other. The development of gunpowder weapons was certainly one of the greatest achievements of the medieval Chinese world.[2] One finds the beginning of it towards the end of the Thang, in the ninth century A.D., when the first reference to the mixing of charcoal, saltpetre (i.e. potassium nitrate), and sulphur is found. This occurs in a Taoist book which strongly recommends alchemists not to mix these substances, especially with the addition of arsenic, because some of those who have done so have had the mixture deflagrate, singe their beards, and burn down the building in which they were working.

After that things happened rather rapidly. The 'fire drug' (*huo yao*), which is the characteristic term for gunpowder mixtures, occurs as igniter in a flame-thrower in A.D. 919 (Fig. 5), and by the time we reach the year 1000 gunpowder packed in simple bombs and grenades was coming into use. Its first composition formulae

[1] In order to complete the pattern of this paper it is necessary however to point out that while the spread of printing in Europe has always been recognized as a necessary precursor of the Renaissance, the Reformation, and the rise of capitalism, because of its democratizing of education, its effects in China were far less. From the Sung onwards the ranks of the scholar-gentry were widely increased by the spread of printing, and the mandarinate was recruited from a much wider circle of families, but the basic structure and principle of the non-hereditary civil service remained essentially quite unchanged. The Chinese social organism had already for centuries been 'democratic' (in the sense of the *carrière ouverte aux talents*) and could therefore absorb a new factor which proved explosive in the aristocratic society of the West. As for transmission, I am satisfied that Gutenberg knew of Chinese movable-block printing, at least by hearsay. *The Invention of [Paper and] Printing in China and their Spread Westwards* is the title of the classical book by T. F. Carter (2nd ed., ed. L. C. Goodrich, New York, 1955).

[2] No study of this in a Western language as yet incorporates all the new knowledge brought to light in recent times by Chinese scholars, especially Fêng Chia-Shêng, but the paper of Wang Ling on 'The Invention and Use of Gunpowder and Firearms in China', *Isis*, 1947, 37, 160, is still broadly speaking correct in its account. For the wider comparative background we now have J. R. Partington's important work: *A History of Greek Fire and Gunpowder* (Cambridge, 1961).

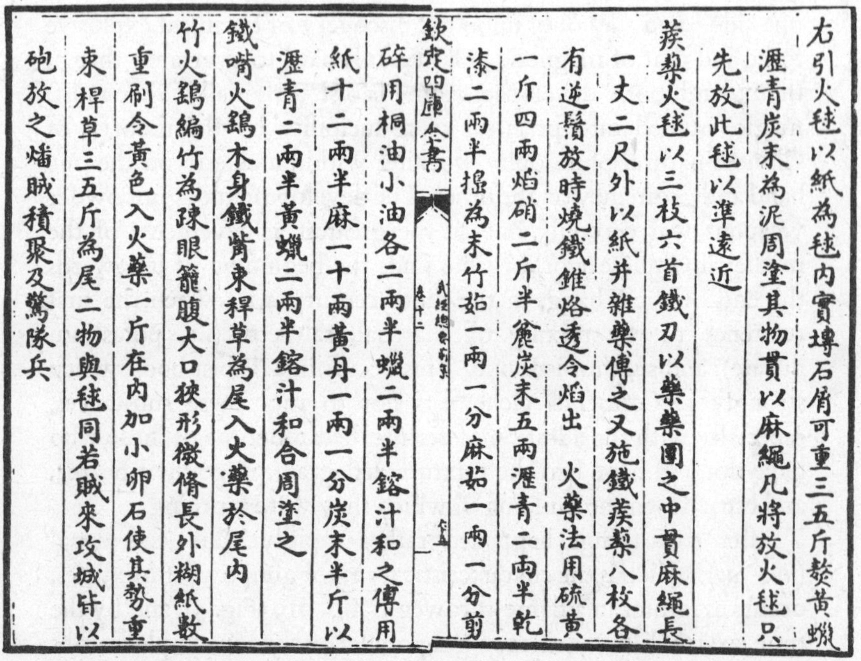

右引火毬以紙為毬內實埄石屑可重三五斤爇黃蠟
先放此毬以準遠近
瀝青炭末為泥固塗其物貫以麻繩凡將放火毬只
蒺藜火毬以三枝六首鐵刃以藥團之中貫麻繩長
一丈二尺外以紙并雜藥傅之又施鐵蒺藜八枚各
有逆鬚放時燒鐵錐烙透令焰出　火藥法用硫黃
一斤四兩焰硝二斤半麄炭末五兩瀝青二兩半乾
漆二兩半擣為末竹茹一兩一分麻茹一兩一分
欽定四庫全書
碎用桐油小油各二兩半蠟二兩半鎔汁和之傅用
紙十二兩半麻一十兩黃丹一兩一分炭末半斤以
瀝青二兩黃蠟二兩鎔汁和合同塗之
鐵嘴火鷂木身鐵背束稈草為尾入火藥於尾內
竹火鷂編竹為疎眼籠腹大口狹形脩長外糊紙數
重剋令黃色入火藥一斤在內加小卵石使其勢重
束稈草三五斤為尾二物與毬同若賊來攻城卽以
砲放之燔賊積聚及驚隊兵

FIGURE 2 Two pages from the *Wu Ching Tsung Yao* of A.D. 1044 showing the earliest gunpowder formula in any civilization. It begins in the sixth column from the right.

appear in 1044 (Fig. 2). This is a good deal earlier than the first references to any gunpowder composition in Europe, 1327, at best 1285. These bombs and grenades of the beginning of the eleventh century did not of course contain a brisant explosive like that which became known in the following two centuries when the proportion of nitrate was raised; they were more like rocket compositions which go off with a 'whoosh' rather than anything which gives a destructive explosion. And indeed it was about this time, the early eleventh century, that a new kind of incendiary arrow (*huo chien*), in fact the rocket, developed. Here immediately we see the importance of the availability of a natural form of tubing, the stem of the bamboo, because it was only necessary to attach a bamboo tube to an arrow and fill it with a low-nitrate composition to get the rocket effect. In this day and age it is hardly necessary to expatiate upon what the Chinese started when they first made rockets fly (Figs. 3, 4).

Thence there followed the important transition to the barrel gun. It occurred early in the twelfth century, about 1120, when the Sung people were conducting their great defensive campaign against the Chin Tartars. In a remarkable book by Chhen Kuei, the *Shou Chhêng Lu*, on the defence of a certain city north of Hankow at that time, there is described the first invention and use of the fire-lance (*huo chhiang*)—a tube filled with rocket composition but not allowed to go loose, held instead upon the end of a spear. An adequate supply of these five-minute flame-throwers, passed on from hand to hand, effectively discouraged enemy troops from storming one's city wall. By about 1230 we begin to have descriptions of really destructive explosions in the later campaigns between the Sung and the Yuan Mongols. Then about 1280 comes the appearance of the metal-barrel gun somewhere in the Old

FIGURE 3 Rocket-arrow and launching box for seventy-five of these projectiles (*Wu Pei Chih*, A.D. 1621).
The rocket is specified as 5 ins. long, and the arrow shaft 2 ft. 3 ins. Such devices date from the beginning of the 11th cent. A.D.

World. As yet we really do not know where it first occurred, whether among the Arabs with their *madfa 'a*, whether among the

Chinese, as seems most likely from the preceding history, or whether possibly among the Westerners. Between 1280 and 1320 is the key period for the appearance of the metal-barrel cannon. I have no doubt that its real ancestry was the substantial bamboo tube of the Chinese fire-lance.

There are two important points to be made about this Chinese development of the first chemical explosive known to man. Firstly, it is not to be regarded as a purely technological achievement. Gunpowder was not the invention of artisans, farmers, or master-masons; it arose from the systematic if obscure investigations of Taoist alchemists. I say 'systematic' most advisedly, for although in the sixth and eighth centuries they had no theories of

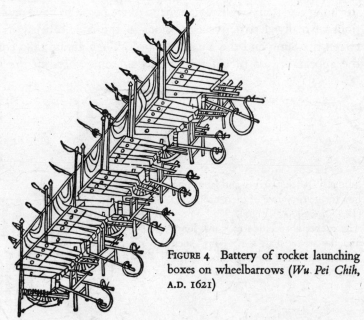

FIGURE 4 Battery of rocket launching boxes on wheelbarrows (*Wu Pei Chih*, A.D. 1621)

modern type to work with, that does not mean that they worked with no theories at all; on the contrary it has been shown that an elaborate doctrine of categories of affinities had grown up by the Thang, reminiscent in some ways of the sympathies and antipathies of the Alexandrian mystical aurificers, but much more developed and much less animistic.[1] I use the term 'mystical aurificers' here because the first alchemists of Hellenistic times, though very interested in counterfeiting gold, and in all kinds of chemical and metallurgical transformations, were not as yet in pursuit of a 'philosopher's stone' which would give a medicine of immortality or an 'elixir of life'. There is every reason for believing that the basic ideas of Chinese alchemy, which had been 'longevity-conscious' from the beginning, made their way to the West through the Arabic world. Indeed, one cannot really speak of

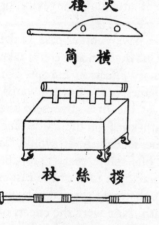

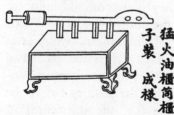

FIGURE 5 'Greek Fire' (*mêng huo yu*) flame-thrower, with tank for the naphtha, and double-acting pump with two pistons to work continuously (*Wu Ching Tsung Yao*, A.D. 1044).
For a reconstruction of the mechanism see *SCC*, Vol. IV, pt. 2, p. 147.

alchemy in the strict sense before the contribution of the Arabs, and it is even claimed that the word itself, and also other alchemical terms, are derived from Chinese originals. Many pieces of chemical apparatus from the Han period have come down to us, such as

[1] See Ho Ping-Yü & J. Needham, 'Theories of Categories in Early Mediaeval Chinese Alchemy', *Journ. Warburg & Courtauld Institutes*, 1959, **22**, 173.

bronze vessels probably used for the sublimation of mercurous chloride (the making of calomel), vapour rising through the two arms and condensing in the centre (Pl. 3). Certain forms of distilling apparatus are also typically Chinese, and quite different from those in use in the West. The distillate, condensed by the vessel of cold water above, drips down into a central receiver and flows out through a side-tube. This is an ancestor of apparatus used in modern chemistry.[1] In sum, the first compounding of an explosive mixture arose in the course of a systematic exploration of the chemical and pharmaceutical properties of a great variety of substances, inspired by the hope of attaining longevity or material immortality.

Secondly, in the gunpowder epic we have another case of the socially devastating discovery which China could somehow take in her stride but which had revolutionary effects in Europe. For decades, indeed for centuries, from Shakespeare's time onwards, European historians have recognized in the first salvoes of the fourteenth-century bombards the death-knell of the castle, and hence of Western military aristocratic feudalism. It would be tedious to enlarge upon this here. In one single year (1449) the artillery train of the king of France, making a tour of the castles still held by the English in Normandy, battered them down one after another at the rate of five a month. Nor were the effects of gunpowder confined to the land; they had profound influence also at sea, for in due time they gave the death-blow to the multi-oared slave-manned galley of the Mediterranean, which was unable to provide gun-platforms sufficiently stable for naval cannonades and broadsides. Less well known, but meriting passing mention here, is the fact that during the century before the appearance of gunpowder in Europe (i.e. the thirteenth) its poliorcetic value had been foreshadowed by another, less lasting development, that of the counterweighted trebuchet, also most dangerous for even the stoutest castle walls. This was an Arabic improvement of the projectile-throwing device (*phao*) most

[1] See Ho Ping-Yü & J. Needham, 'The Laboratory Equipment of the Early Mediaeval Chinese Alchemists', *Ambix*, 1959, **7**, 58.

characteristic of Chinese military art, not the torsion or spring devices of Alexandrian or Byzantine catapults, but the simpler swape-like lever bearing a sling at the end of its longer arm and operated by manned ropes attached to the end of its shorter one.

Here the contrast with China is particularly noteworthy. The basic structure of bureaucratic feudalism remained after five centuries or so of gunpowder weapons just about the same as it had been before the invention had developed. The birth of chemical warfare had occurred in the Thang but it did not find wide military use before the Sung, and its real proving-grounds were the wars between the Sung empire, the Chin Tartars, and the Mongols in the eleventh to thirteenth centuries. There are plenty of examples of its use by the forces of agrarian rebellions, and it was employed at sea as well as on land, in siege warfare no less than in the field. But as there was no heavily armoured knightly cavalry in China, nor any aristocratic or manorial fuedal castles either, the new weapon simply supplemented those which had been in use before, and produced no perceptible effect upon the age-old civil and military bureaucratic apparatus, which each new foreign conqueror had to take over and use in his turn.

IV

Next let us look at the third of Bacon's great discoveries. If Ptolemaic astronomy was purely Greek, the early study of magnetism was purely Chinese, a point of immense importance. If we go into any place today where nature is under accurate observation or control—into an atomic power-station, the engine-room of an ocean liner, or any scientific laboratory—the walls are covered with dials and pointers, and people are making dial-and-pointer readings. But the first of all dial-and-pointer devices, so classical in the philosophy of science, was the magnetic compass, and in the development of this Europe had no part.

In the Sung period we find as one of its early forms a small piece of lodestone embedded in the body of a wooden fish with a little needle projecting from it; floating in water, it indicates the south.[1] The same thing was done with a dry suspension; a thin chopstick cut off and sharpened to a point bore the lodestone inside a little wooden turtle, with again a needle sticking out to add a small amount of extra torque. These designs date from about 1130, but we have a still earlier one, from 1044, described in a book called *Wu Ching Tsung Yao* (Compendium of Important Military Techniques) by Tsêng Kung-Liang. This is nothing other than the 'floating fish' so often mentioned by Arabic writers later on, the cup-shaped fish of magnetized iron floating on the water.

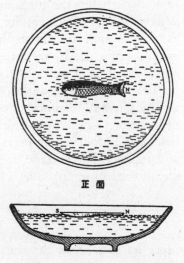

FIGURE 6 Floating iron compass using remanent magnetism (*Wu Ching Tsung Yao*, A.D. 1044, reconstruction by Wang Chen-To).

Still more interesting, this compass-fish was not magnetized by being rubbed on the lodestone but by being heated to red heat while held in a north–south position in the earth's magnetic field. Remanent magnetism is a surprise to meet with in the early eleventh century. By its end, the most usual thing was to have a magnetized needle suspended on a single thread of raw silk.

To get back to the beginning one has to mention the divination devices called *shih*. Used by the diviners in Han times, they had a

[1] We are, of course, accustomed to think of the needle as pointing to the north, but in China the south was always considered to be indicated. In Chinese cosmic symbolism the emperor represented the pole-star, and so faced south on his throne, theoretically doing nothing, yet ruling all things with perfect success.

square earth-plate surmounted by a discoidal heaven-plate, having the Great Bear carved on its upper surface together with the standard cyclical characters, compass points, lunar mansions, constellation names, and so on. In Wang Chhung's *Lun Hêng* (Discourses Weighed in the Balance), written in A.D. 83, there is a text which says that if you take the 'south-controlling spoon' (*ssu-nan chih shao*), and throw it on the ground, it will always point to the south. The accepted view is that to 'throw it on the ground' was not meant literally, but indicated that it was to be placed on the ground-plate of the diviner's board. The spoon itself was an actual piece of lodestone, carved into the shape of the Northern Dipper (the Great Bear), i.e. into a Chinese spoon. It has been found experimentally that this can in fact be done if the bronze plate is polished as highly as possible; then the torque will rotate the spoon so that it turns to the south. Originally the spoon had been but one of a number of magic models of the heavenly bodies used in techniques of divination allied to board games (Pl. 4).

It is true that this device is a reconstruction from a text, and that an actual spoon made of lodestone has not so far been found in any tomb. But during the following thousand years there are constant literary references to a 'south-pointer' which can only be explained if something of this kind existed. Later on there is a firm priority of two or three centuries before the first European mention of magnetic polarity about 1180. Indeed it is true to say that people in China were worrying about the declination[1] before Europeans even knew of the polarity. It is a remarkable fact that past variations of declination can be found embalmed in the Chinese geomantic compass. This has three circles, not only for the astronomical north–south, but another with all the points staggered $7\frac{1}{2}°$ east of it and another similarly $7\frac{1}{2}°$ west. Thus the geomantic compass preserves a record of declinations which at certain times were east of astronomical north–south, and then west.

The priority in knowledge of polarity, induction, remanence, declination, etc., was also maintained in priority of first use in

[1] That is to say, the variable deviation of the magnetic needle from astronomical north.

navigation, which must have started at least as early as the tenth century. We have charts of the early fifteenth century which show itineraries diagrammatically like steamship routes across the oceans, with the compass-bearings marked along them. So many watches on such-and-such a bearing, then change course and carry on again for a prescribed time, etc. Such knowledge came to the West, but how it came remains a mystery. Perhaps some Arabic or Indian text will throw light upon it. Perhaps the knowledge even travelled overland through Tartar kingdoms and not by sea at all.

Magnetical science was indeed an essential component of modern science. All the preparation for Peter of Maricourt, the greatest medieval student of the compass, and hence for the ideas of Gilbert and Kepler on the cosmic role of magnetism, had been Chinese. Gilbert thought that all heavenly motions were due to the magnetic powers of the heavenly bodies, and Kepler had the idea that gravitation must be something like magnetic attraction. The tendency of bodies to fall to the ground was explained by the idea that the earth was like an enormous magnet drawing things unto itself. The conception of a parallelism between gravity and magnetism was a vitally important part of the preparation for Isaac Newton. In the Newtonian synthesis gravitation was axiomatic, one might almost say, and spread throughout all space just as magnetic force would act across space with no obvious intermediation. Thus the ancient Chinese ideas of action at a distance[1] were a very important part of the preparation for Newton through Gilbert and Kepler. The field physics of still later times, established in Clerk Maxwell's classical equations, and more congruent with organic thought than Greek atomic materialism, can again be traced back to the same root. Hence the concluding words, entirely justified, of the passage quoted above on p. 64.

Mutatis mutandis one can make the same two statements about

---

[1] A valuable history of this conception in relation to that of continuous contact action (though only so far as Western thought is concerned) has been written by Mary Hesse: *Forces and Fields: the Concept of Action at a Distance in the History of Physics* (London, 1961).

the magnetic compass as were made about gunpowder. It was not a purely empirical or technological achievement because the Taoist geomancers had their theories during its long developmental period, as we know well from many texts that have been preserved. The fact that these theories were not of the modern type does not entitle us to ignore them. The whole discovery had arisen from a divination procedure or cosmical magic, but what carried it forward was the Chinese attachment to a doctrine of action at a distance, or wave-motion through a continuum, rather than direct mechanical impulsion of particles; atomism being foreign to them, this it was which led them on to see nothing impossible in the pole-pointing property of a stone or of iron which had touched it. Secondly, the magnetic compass, or more broadly speaking, the knowledge of magnetic polarity as well as magnetic attraction, had also its sociologically earth-shaking character in the Western world. The part which it played in the nascent phases of modern science would be sufficient justification for this in itself, but there was more; for in the hands of the European sea-captains of the fifteenth century the compass crowned a whole period of navigational science which had been inaugurated in the thirteenth, and made possible not only the circumnavigation of the African, but the discovery of the American, continent. How profoundly this affected the life of Europe, with the influx of vast quantities of silver, the marketing of innumerable new kinds of commodities, and the opening up of colonies and plantations, hardly requires elaborate emphasis here, when even elementary textbooks tell the story. But again there is the other side of the picture. Chinese society was not upset by the knowledge of magnetic phenomena; the geomancers continued to advise families upon the best siting of houses and tombs with ever-increasing refinement of their baseless art,[1] and the sea-captains

---

[1] I say 'baseless', because the idea that good or evil fortune would follow the proper situation of dwellings or tombs was purely proto-scientific, or as some would say, superstitious. But one must not forget that a very strong aesthetic element entered into medieval Chinese geomancy, as is evidenced still today by the exquisite patterns in which farmhouses, paths, towns, pagodas, and all kinds of human habitations blend with the physiographic scenery.

continued to find their way to the East Indies or the Persian Gulf in a trade that was peripheral to China's main economic life.[1]

V

We may now leave the thoughts which flood in upon the reader of Francis Bacon's passage, and go on to consider a number of other outstanding scientific and technological gifts of China to the world. The scientific material which I have chosen may be divided into three parts: (a) explosives chemistry or proto-chemistry; (b) magnetic physics and the mariner's compass; (c) astronomical co-ordinates and instruments, mechanical clockwork, and the 'open' cosmology. Having discussed the first two already, we shall now deal with the third. Afterwards there will follow four technological subjects: (a) the use of animal-power, with the inventions of the stirrup, the efficient equine harnesses, and the wheelbarrow; (b) the use of water-power, with associated inventions such as the driving-belt, the chain-drive, the crank, and the morphology of the steam-engine; (c) iron and steel technology, bridge-building, and deep drilling; (d) nautical inventions such as the stern-post rudder, fore-and-aft sailing, the paddle-wheel boat, and watertight compartments. It must be emphasized that these are only a selection from a large variety of choice, and that the selection made is particularly deficient on the biological side.[2] It will also be worth

[1] Even the great period of maritime expansion during the first half of the fifteenth century, when the fleets of the Ming navy under the admiral Chêng Ho repeatedly found their way as far as Madagascar, Medina, and Muscat, to say nothing of the spice islands and the northern fur coasts, had little effect upon Chinese economic life as a whole, and certainly never ran any risk of switching it into some new track.

[2] For example there is no room to say anything of physical meteorology or mineralogy among the inorganic sciences; or of sphygmology, nutritional science, entomology, plant protection, etc., among the biological ones. It is nevertheless certain that the study of the pulse and the empirical discovery of deficiency diseases in China influenced general scientific thinking from the seventeenth down to the end of the nineteenth centuries. As for technology, we have no room to say anything of the ceramics industry or the porcelain which Europeans strove so much to imitate in the eighteenth century, nor of the first plastic, lacquer, nor of mines, nor yet of fisheries.

while to consider the chronologcial order in which the transmissions to Europe occurred, coming as it were in 'clusters' at particular times rather than one by one over long periods of time. Finally, I shall return to the paradox already adumbrated of European social instability compared with Chinese stability, and link it with that other paradox of the primary Asian success in applying science to human needs, followed by the secondary European success in discovering the method of scientific discovery itself and so inaugurating modern as opposed to mediaeval science and technology.

There are three ways of measuring the position of any star in the heavens, and modern astronomy uses, not the ecliptic coordinates of the Greeks or the altazimuth measurements of the Arabs, but the equatorial system of the Chinese (Fig. 7). The measurement of position on the surface of the celestial sphere (the apparent dome of the heavens) was accomplished in all civilizations by building graduated circles into an armillary sphere. The greatest Hellenistic astronomer Ptolemy (second century A.D.) had such an instrument at his disposal, and it lives on in the location gear of the modern telescope, for the latter is simply a sighting-tube of vastly increased size and power, not a finding mechanism. The sighting-tube and the graduated rings were the two essential elements for ascertaining celestial positions.[1]

Now the time of the development of the armillary sphere in China and in Greece is about the same, if indeed the Chinese do not have some priority. It is fully present by the time of Chang Hêng, that great scholar and scientist of the Later Han whose period of activity was from A.D. 100 to 130, just before Ptolemy. But it is quite likely that it was already complete in most of its details as early as the time of Lohsia Hung, who was repairing the calendar about 100 B.C. in the Former Han period; and rings of some kind must have been used by Shih Shen and Kan Tê in about 350 B.C. if the tradition is right that they were the first to give

---

[1] The Chinese did not use the 360° graduation, but one of $365\frac{1}{4}°$, based on the number of days in the year. While at first sight this seems very awkward, it had some concrete advantages.

star positions in degrees. Even in these early times measurements were always equatorial. One of the finest Chinese instruments was the armillary sphere of Su Sung, set up in 1088 at Khaifêng, the capital of the Northern Sung Dynasty. This was the first observational instrument in astronomical history to be provided with a clock-drive. The finest extant Chinese instrument is no doubt the bronze armillary sphere of Kuo Shou-Ching, the great Yuan astronomer who re-equipped the observatory at Peking in 1275. It is now at Nanking in the grounds of the Purple Mountain Observatory.

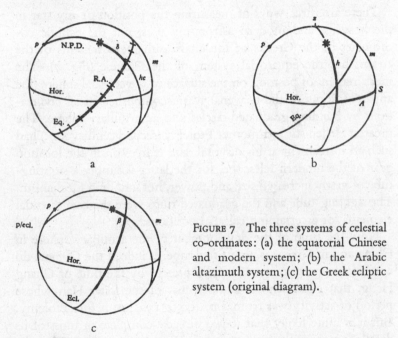

FIGURE 7 The three systems of celestial co-ordinates: (a) the equatorial Chinese and modern system; (b) the Arabic altazimuth system; (c) the Greek ecliptic system (original diagram).

If the sighting-tube was destined to swell and the graduated circles to shrink, the line of progress lay in dissecting the concentric cage of the armillary sphere. The equatorial mounting of the modern telescope was invented in China three and a half centuries before any telescopes existed. If one takes all the concentric circles apart and mounts them non-concentrically in their correct planes suitably connected together, one has an instrument which came later to be called 'Turkish', i.e. the 'torquetum'. Its first inventor, the Spanish Muslim Jābir ibn Aflah, designed it in the twelfth century largely as a kind of computing machine for transferring from one set of co-ordinates to the others. But when it was introduced to China by the scientific mission of Jamāl al-Dīn in 1267 it quickly led Kuo Shou-Ching to the invention of a device called *chien i* or 'simplified instrument'. This was essentially the torquetum with the ecliptic components omitted, and it

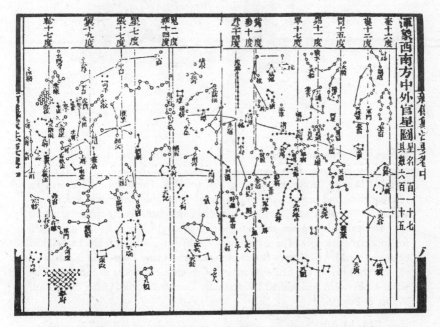

FIGURE 8 Star-chart from Su Sung's *Hsin I Hsiang Fa Yao* (A.D. 1094) showing fourteen of the twenty-eight *hsiu* (lunar mansions), with many of the Chinese constellations contained in them. The equator is marked by the central horizontal line; the ecliptic arches upward above it. The legend on the right-hand side reads: 'Map of the asterisms north and south of the equator in the SW. part of the heavens, as shown on our celestial globe; 615 stars in 117 constellations.' The *hsiu*, reading from the right, are: Khuei, Lou, Wei, Mao, Pi, Tshui, Shen (Orion), Ching, Kuei, Liu, Hsing, Chang, I, and Chen. The unequal equatorial extensions are well seen.

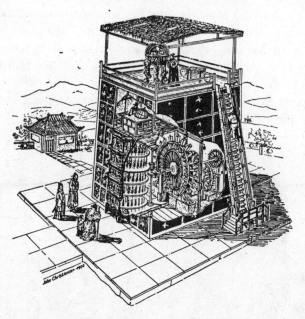

FIGURE 9 Reconstruction by John Christiansen of the great astronomical clock erected by Su Sung, Han Kung-Lien, and their collaborators at the Imperial Palace, Khaifêng, Honan province, in A.D. 1088–92. The water-wheel rotated a celestial globe and an armillary sphere, and also an elaborate series of jack figures which announced the time without any dial clock-face. As the sphere was used for observational purposes, this was the first of all clock-drives such as are used for modern telescopes. The escapement for hydro-mechanical clockwork had been invented by I-Hsing (a Buddhist monk) and Liang Ling-Tsan in A.D. 725, six centuries before the first appearance of mechanical clocks in the west. The water-tanks were replenished by manually operated norias. (After Needham, Wang & Price).

was indeed the forerunner of the mounting of all modern telescopes (Pl. 5).

It is an extraordinary fact in the history of science that the Chinese were able to make such brilliant advances, pushing far ahead of the West (except for the complicated astrolabe), without full knowledge of geometry in its deductive Euclidean form. In any case it was the father of modern observational astronomy, Tycho Brahe, who in the sixteenth century introduced both the

Chinese practices, the equatorial mounting and the equatorial co-ordinates, into modern science, which has never since departed from them. His explicit reason was the greater instrumental accuracy, but he possessed Arabic astronomical books, and the Arabs knew well what the Chinese usage was.

As has just been mentioned, the Khaifêng armillary sphere of 1088 was provided with a clock-drive. How was this possible? It was because China was responsible for the development of the mechanical clock, not Europe; a story which has only very recently been brought to light.[1] Indeed the mechanical clocks of China built between A.D. 700 and 1300 have revealed at last the missing link between the very ancient water-receiving and water-giving vessels (clepsydras) of Babylonia and ancient Egypt, and the purely mechanical clocks and watches of later ages. From the time of Chang Hêng onwards the Chinese were interested in making uranographic models (celestial globes, etc.) revolve by water-power, and it was precisely their equatorial preference which suggested the idea to them. In the earliest form of European mechanical clock (from 1300 onwards), the verge-and-foliot escapement is what dissects the passage of time, the two pallets of the verge alternately arresting the rotation of the crown-wheel, which is powered by a falling weight. In this way the familiar tick-tock movement is achieved, and the whole system is slowed down to the rate of the apparent revolution of the heavens, which is of course man's primary clock. But for six hundred years before this another kind of mechanical clock had been in existence, though only in the Chinese culture-area.

We may take as an example the instrument described in a book entitled *Hsin I Hsiang Fa Yao* and written by Su Sung in 1092 about the great clock-tower which had been put up in Khaifêng a few years before. A general reconstruction can be described as follows. The machinery is inside the building on the right, and the time-

---

[1] For a full account see *Heavenly Clockwork* by J. Needham, L. Wang, & D. J. de S. Price (Cambridge, 1960) (Antiquarian Horological Society Monograph, no. 1). More briefly, J. Needham, 'The Missing Link in Horological History; a Chinese Contribution', *Proc. Roy. Soc. A.*, 1959, **250**, 147 (Wilkins Lecture).

telling apparatus is on the left, with the puppets in their pagoda coming round to announce the hour, and ringing bells and gongs. Above the time-annunciator system one can see the celestial globe which rotated automatically, and lastly on the roof is the armillary sphere which also rotated automatically. The main drive was not a falling weight but a water-wheel. One can see, too, the wheels behind it for getting the water up into the reservoir again. The essential part of the time-keeping was a linkwork escapement quite different from the verge-and-foliot system. Water poured continually from a constant-level tank into the scoops of the water-wheel, but each one could not go down until it was full. As it went down it tripped a couple of levers or weighbridges which by linkwork connexions released a gate at the top of the wheel and let it move on by one scoop. One might say that the machine was arranged so as to dissect time by the accurate and rapid weighing of successive small quantities of a fluid.[1] The main driving-wheel rotated a driving shaft which powered all the puppet wheels, the celestial globe, and also the armillary sphere. In later developments, Mark II and Mark III as one might say, the vertical shaft was replaced by a chain-drive, almost certainly the oldest power-transmitting chain-drive known to history. Working models have been built of the water-wheel linkwork escapement which keep good time.[2]

The Chinese hydro-mechanical clock thus bridges the gap between the clepsydra and the weight- or spring-driven clock. It was not entirely dependent on the constant flow of a liquid, as the clepsydra had been, because its time-keeping properties could be adjusted by varying the counter-weights on the weighbridges. As for its first origin, one finds this in a remarkable clock built by the Tantric Buddhist monk I-Hsing and an engineer Liang Ling-Tsan for the Thang court at the College of All Sages (one of the

---

[1] In this case it was water but in some other medieval Chinese clocks it was mercury, which does not freeze.

[2] For example by our collaborator Mr. John Combridge, who demonstrated one first at the History of Science Symposium at Worcester College, Oxford, and then at the London Planetarium Reception for the Scientific Delegation of Academia Sinica, in the summer and autumn of 1961.

imperial institutes of higher learning) about A.D. 725.[1] The tradition lasted on well into the Ming and had still not died out by the time of the Jesuit mission in the seventeenth century, a thousand years later, when it was replaced by the more compact and practical Renaissance clockwork (cf. Fig. 31 on p. 278).

What was the significance of all this for Europe? The astronomical achievements had of course no direct social consequences; they were simply incorporated into the body of modern astronomy, with all that that implied for the profound changes in world-outlook which have taken place since the seventeenth century. Thus to the break-up of the naïve cosmology of medieval Christendom, seen still in Dante, Chinese influence indirectly contributed. We shall see in a moment how it contributed directly. Clockwork had more obvious and immediate effects. Although the details of any transmission are still obscure there are good grounds for thinking that the Chinese water-wheel linkwork escapement was known and used in thirteenth-century Europe; at the least there was knowledge that the problem of mechanical time-keeping had in principle been solved. From its first European beginnings clockwork generated a type of craftsmanship which together with that of the millwrights was vitally important for the development of mechanical and industrial production in the post-Renaissance period. Moreover the mechanical clock excited Europeans because it embodied the properties of the cosmic models from which it had originated. As Lynn White says in his recent admirable book on the history of technology:

'Suddenly, towards the middle of the fourteenth century, the mechanical clock seized the imagination of our ancestors. Something of the civic pride which earlier had expended itself in cathedral-building was now diverted to the construction of astronomical clocks of outstanding intricacy and elaboration. No European community felt able to hold up its head unless in its midst the planets

[1] This estimate is based on philological grounds—the similarity of the technical terminology used. But the invention may well go back a good many centuries earlier if the terminology radically changed, since we have numerous descriptions of celestial globes which say that they were accurately rotated by water-power, but do not give details of the mechanism.

wheeled in cycles and epicycles, while angels trumpeted, cocks crew, and apostles, kings and prophets marched and countermarched at the booming of the hours.'[1]

Thus the mechanical orreries which had long graced the courts of Chinese emperors and princes entered the service of those European city-states which were soon to burst through the bonds of the feudalism which surrounded them. At the same time the urano-graphic models succeeded to the inheritance of Ptolemaic planetary astronomy, which constituted a further stimulating challenge to mechanization. Conversely the powered models soon became symbols of the implicit tendencies of the scientific Renaissance. The explanation of Nature in terms of the 'analogy of mechanism' was one of the most fundamental concepts which led to the success of modern science, displacing the older analogies derived from organic growth, sympathies and antipathies, or human techniques.[2] Lynn White goes on to point out that at the time when the problem of mechanical time-keeping was first solved in Europe a new theory of impetus was emerging, transitional between that of Aristotle and the inertial motion of Newton. Now

'regularity, mathematically predictable relationships, facts quantitatively measurable, were looming larger in men's picture of the universe. And the great clock, partly because its inexorability was so playfully masked, its mechanism so humanized by its whimsicalities, furnished the picture. It is in the works of the great ecclesiastic and mathematician Nicholas Oresmus, who died in 1382 as Bishop of Lisieux, that we first find the metaphor of the universe as a vast mechanical clock created and set running by God so that 'all the wheels move as harmoniously as possible'. It was a notion with a future; eventually the metaphor became a metaphysics.'[3]

No more needs to be said, but when to all this one adds the simple fact that the measurement of time is one of the handful of abso-lutely indispensable tools of modern science, it can be seen that I-Hsing and Su Sung started something.

It remains to justify the statement that China provided a direct contribution to the modernization of the world-picture. Passing reference has already been made to this (p. 23). In brief, the medi-

[1] *Mediaeval Technology and Social Change* (Oxford, 1962), p. 124.
[2] For a recent and lucid account of this question see Mary Hesse, op. cit., pp. 30 ff.
[3] Lynn White, op. cit., p. 125.

aeval cosmology of the Chinese (including Buddhist trends) was far more 'open' than that of mediaeval Europeans. There were in China three classical astronomical cosmologies: the archaic *kai thien* sky-dome (allied to still older Babylonian conceptions), the normal doctrine of the celestial sphere (*hun thien*), which did not commit itself to the nature of the phenomena beyond their geometrical relationship, and thirdly the *hsüan yeh* theory, for which the stars and planets were lights of unknown substance floating in infinite empty space. This last view was the one most commonly held by Chinese astronomers in historical times, and it chimed in well enough with the infinities of time and space, both great and small, postulated by Buddhist scientific thinkers. It took untold time for an object thrown from one Buddhist heaven to reach another, or to fall to earth; and Thang calculations of the eighth century A.D. cheerfully fixed ancient astronomical events a hundred million years before that time—much in contrast with the eighteenth-century European bishop's estimate of the date of creation as 22 October, 4004 B.C. at six o'clock in the evening. Chinese astronomy had always been equatorial and diurnal, not ecliptic and annual, so that it had little of that planetary astronomy for which the Greeks had needed Euclid, but on the other hand this brought some compensating advantages—the Chinese never became enamoured of the circle as the most perfect of all geometrical figures, and hence were never the prisoners of the concentric crystalline celestial spheres which Westerners had found necessary to explain the motion of the planets and the apparent rotation of the fixed stars. Hence their influence was a liberating one when Europeans were breaking forth from this prison. Whether any breath of it reached men such as Giordano Bruno and William Gilbert, who attacked the Ptolemaic–Aristotelian spheres before the end of the sixteenth century, we do not know, but it is quite sure that fifty years later European thinkers who were adopting Copernicanism and abandoning the spheres drew much encouragement from the knowledge that the wise astronomers of China (Europe's sinophile period was just beginning) had never had any use for them.

85

VI

It is now time to descend from these high celestial regions and to pay attention to some of the more workaday techniques which Chinese ingenuity contributed to the rest of the world—the stirrup, the efficient equine harnesses, and the simple wheel-barrow. About the foot-stirrup there has been a great deal of discussion, and after it had been attributed on what seemed excellent evidence to the Scythians, the Lithuanians, and especially the Avars, recent critical analysis has ruled in favour of China.[1] Tomb-figures from the Chin Dynasty (A.D. 265–420) show it clearly (Pl. 6), and the first textual description comes from very little later (477), after which representations become numerous. The stimulus was no doubt Indian, mediated through Buddhist contacts, rather than nomadic, for the toe-stirrup (only useful for unshod riders in a hot climate) appears in sculptures at Sanchi and elsewhere in the second century B.C. Foot-stirrups did not appear in the West (or Byzantium) until the early eighth century,[2] but their sociological influence there was quite extra-ordinary. 'Few inventions,' says Lynn White, 'have been so simple as the stirrup, but few have had so catalytic an influence on history.'[3] It effected nothing less than the application of animal-power to shock combat. The cavalryman was welded into a unit with his steed in a way which none of the Asian mounted archers had ever been, so that he had only to guide, rather than to deliver,

[1] See the brilliant and well-documented study of Lynn White, op. cit., pp. 2, 14 ff., 28 ff. I cannot accept, however, his cavalier dismissal of the evidence of the Wu Liang tomb-shrines (p. 141) dating from A.D. 147; this date is not contested by sinologists and everything depends on the credibility of the rubbings made by the Fêng brothers in 1821, for much weathering has occurred since their time. In any case this question does not affect the general argument.

[2] As in many other transmissions the means whereby the invention came remain completely unknown. In all such cases the burden of proof clearly lies at the door of those who would wish to maintain independent invention, and the longer the period elapsing between two appearances of an invention the less likely independent invention is.

[3] Op. cit., p. 38.

the blow. Horsemen fighting in this new manner with the Carolingian wing-spear, and gradually more and more enveloped in protective metal armour, came in fact to constitute the familiar feudal chivalry of nearly ten European medieval centuries. It may thus be said that just as Chinese gunpowder helped to shatter European feudalism at the end of this period, Chinese stirrups had originally helped to set it up.

A more intractable problem is why nothing of this kind happened in China. Once again we face the astonishing stability of that civilization. So deeply civilian was its ethos that the very conception of aristocratic chivalry was perhaps impossible. Perhaps if the invention had come in the feudal period of the Warring States, before bureaucratism really settled into the saddle, the story might have been very different. It may be that the tradition of the mounted archer, adopted already in China as early as the fourth century B.C., was too strong to be overcome. Perhaps after all this tradition was fundamentally superior in military science, for when the Mongolian cavalry came at last face to face with the armoured knights of medieval Europe in the thirteenth century, it was not the knights who won the day in any battle; the withdrawal of the Mongols from the West was due to internal political events rather than occidental resistance.

Besides originating the foot-stirrup, China was the only ancient civilization in which the linkwork problem of efficient harness for equine animals was solved.[1] Here again the consequences were almost incalculable. Harnessing the bovine animal is comparatively easy because the ox is of a very convenient anatomical shape. The cervical vertebrae rise in a sort of hump which enables the yoke to bear against them. But this will not work for horses, donkeys, mules, or any equine animal, such vertebral projections being absent. Throughout the centuries there have been only three main ways of dealing with this. The so-called throat-and-

[1] I can never touch upon these problems without paying a tribute to that remarkable man Lefebvre des Noëttes, who in his classical work *L'Attelage et le cheval de selle à travers les âges* (Paris, 1931) first posed the question of the history of harness and its social consequences.

girth harness was characteristic of antiquity all over the Old World and lasted in Europe down to as late as the fifth or sixth century A.D. At the other end of the scale is the modern collar harness, a hard part and a soft part being combined in one, and so arranged that the pull comes from the sternal region of the horse. In the throat-and-girth harness, on the contrary, it came from the back of the horse, occluding the trachea and half-choking the animal, which therefore could not exert more than a quarter or a third of its tractive power. With collar harness whether in traces or shafts it can pull thoroughly well. But there is an alternative way of accomplishing this desirable objective, namely to use the breast-strap harness, where a trace, suspended by a withers strap, surrounds the animal, thus pulling also from the sternum.

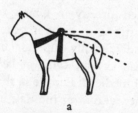

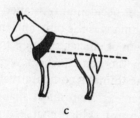

FIGURE 10 The three main systems of equine harness: (a) the throat-and-girth harness characteristic of occidental antiquity; (b) the efficient breast-strap harness of ancient China; (c) the collar harness first developed in early mediaeval China (original diagram).

The dates of these harness forms are of course vitally important. Ancient Egyptian carvings and Greek vase-paintings always show the typical throat-and-girth harness; Roman sources also.[1] The first representation of the breast-strap harness in Europe occurs on

[1] The Hellenistic and Gallic Romans seem however to have experimented with other forms of harness, most of which did not survive. Ancient Indian harness forms are also obscure. These complicated questions will be found discussed in SCC, Vol. IV, pt. 2, sect. 27 f.

an Irish monument of the eighth century, though there is linguistic evidence that it was known among Slavs and Germans a couple of centuries earlier. But in China we find this very much earlier still. Some time between the Shang period (c. 1500–1027 B.C.) and the Chhin unification (third century B.C.), probably during the early Warring States period, the breast-strap harness came into universal use, and one sees it invariably in the carvings and moulded bricks of the Han Dynasty (206 B.C.–A.D. 220). The Wu Liang tomb-shrine reliefs of c. A.D. 147 show the breast-strap harness on the chariots of the two secretaries and historians observing the famous incident of the Battle on the Bridge. Collar harness in Europe first appears early in the tenth century, as we know from Frankish miniature paintings. But here again China has precedence, for in a magnificent fresco of 851, the triumphal procession of the Exarch of Tunhuang, in the 'Thousand-Buddha Caves', one sees the harness clearly on five horses in the shafts of a series of carts in the train of the exarch's consort (Pl. 7). In careful copy enlargements one can see the soft padded collar and a yoke-like cross-bar resting upon it between the shafts. So the collar was essentially a soft cushion devised to replace the 'hump' of the ox and to allow a shaped cross-bar to be placed against it. The oldest pictures of horses and carts at the Chhien-fo-tung cave-temples date from c. 485 to 520, and although they do not show the collar itself the arrangement is quite clear because the pull is coming from the sternal region, and without a collar the 'yoke' would not have stayed in position at all. Its presence may therefore be inferred with confidence. A throat-and-girth arrangement is out of the question partly because the Chinese had already abandoned it for some eight hundred years, and partly because it never occurs in any civilization combined with shafts.[1] Breast-strap harness is also excluded partly because no such straps are to be seen, while the hard part of the collar is. I consider therefore that these late fifth-century and early sixth-century pictures give unambiguous evidence for collar harness, and between this date and the mid-

[1] Never normally, that is, but it may perhaps have been tried in some of the Roman experiments just mentioned.

ninth century there are many more fresco paintings at Chhien-fo-tung of the same kind.[1] Particularly interesting is the fact that the collar harness used today in Kansu province and all over North China still consists of two parts, the annular cushion (*tien-tzu*) and a kind of wooden framework in front, the *chia-pan-tzu*, which is of course a development of the old cross-bar 'yoke', and is attached by cords to the ends of the shafts. One finds two-part collar harness also in other parts of the world, for example in Spain, where perhaps it is a relict form of what was brought by the Arabs. As for the first origin of the annular cushion, philological evidence indicates that it was taken from the pack-saddle of the Bactrian camel.

When we pass from the archaeological origins of efficient equine harness to the panorama of effects which its introduction to the West brought about, we enter a realm already long worked over by occidental historians, and generally regarded by them as of the highest importance for the development of feudal (and ultimately capitalist) institutions. Drawing again from a recent and penetrating study of Western medieval technology, we may say that the general adoption of the heavy plough in northern Europe was only the first stage in the agricultural evolution of the early Middle Ages; the next thing was to acquire such harness as would make the horse an economic as well as a military asset.[2] The horse exerts no greater tractive force than the ox, but its natural speed is so much faster that it produces fifty per cent more foot-pounds of energy per second; moreover it has greater endurance and can work one or two hours longer each day. But although Chinese breast-strap harness had been available from about 700 (if not by 500 in eastern Europe) and Chinese collar harness from about 900, the harnessing of the horse to the plough

[1] The Tunhuang material has been fully discussed and figured by J. Needham & Lu Gwei-Djen, 'Efficient Equine Harness; the Chinese Inventions', *Physis*, 1960, **2**, 143.

[2] Lynn White, op. cit., pp. 57 ff., 61 ff., 67 ff. I regret, however, that I cannot accept his earliest datings for the collar harness in Europe, nor his interpretations of the Tunhuang frescoes, the Oseberg tapestry, the Swedish 'horse-collars', and other things. My views are set forth in *Science and Civilisation in China*, loc. cit.

was slow in coming. About 860 King Alfred heard with surprise from Ohthere that in Norway what little he ploughed he ploughed with horses, but no pictorial evidence is available until the Bayeux Tapestry (c. 1080), by which time a number of texts confirm it. Efficient equine harness was associated with many changes, including crop rotation and a great improvement of the nutritional level of man and beast, but here we can only look at two social effects. One was a marked decrease in the expense of land haulage so that cash-crop produce could travel far more effectively than before; and a considerable technical development in transport vehicles, notably four-wheeled wagons and carriages with improved pivoting front axles, brakes, and eventually springs. The other was more sociological, a proto-urbanization of rural settlement. Since the horse could move so much faster than the ox, the peasant no longer had to live in close proximity to his fields, and thus villages grew at the expense of hamlets, small towns at the expense of villages. Naturally life was more attractive in the larger units; they were more defensible, they could support bigger and better churches, schools, and inns, and commercial facilities penetrated more easily to them. If they grew enough they might hope for a charter. They were in fact the precursors of those urban units later on to be paramount in European culture. Thus by an extraordinary paradox the inventions of a feudal-bureaucratic civilization to which the city-state conception was quite foreign reinforced the intrinsic tendencies within Western feudalism towards a city-state culture which would in time generate an entirely new form of social order.

Why did none of these effects happen in China? In the first place there were no city-state traditions, and any tendency to agglomeration only gave rise to another administrative centre held for the emperor by civil and military authority. More important, in at least half the country neither ox nor horse but the water-buffalo was the essential plough animal, and there was no substitute for it in wet rice cultivation until today's petrol tillers. The entire agricultural picture was so different that horse harness could not affect it in the same way. Transport over land it did affect, but relatively little because China, at any rate since the Han period, had depended for communications primarily on rivers and canals. In military affairs also the canals and irrigation ditches made the Chinese countryside unsuitable for cavalry warfare, as many a nomadic leader from the Tobas to the Mongols found out to his cost. The horse was thus at a disadvantage in Chinese conditions, and though always a factor to be reckoned with, could not affect the life of the culture as profoundly as it did in Europe.

We can dispose of the wheelbarrow in a single paragraph. No pictorial or other evidence of it is known in Europe before the

Boratu

FIGURE 11 Gerard Mercator's representation of a Chinese land-sailing carriage (A.D. 1613).

Chinarum gens admodum ingeni-
osa esse perhibetur, adeo ut currus
excogitarint fabricaverintque,
quos velis ventisque per campos
et loca plana, uti navigia per
mare derigere optime norint.

thirteenth century, at which time it doubtless played its part in the building of the great medieval cathedrals. In China, however, it is associated with the famous Shu general of the Three Kingdoms period, Chuko Liang (third century A.D.), who used it for supplying his armies; but philological evidence of some weight takes the wheelbarrow back to the middle of the Han period, i.e. the beginning of our era. The replacement of one pair of hands at the end of a hod or stretcher by a wheel would seem to be a piece of mechanization so absurdly simple that all civilizations would have had it from the earliest times, but this is not the case. Nor is the implicit picture of its evolution justified, for the wheel of the Chinese wheelbarrow was not most typically at one end, it was rather in the centre, thus suggesting that the invention was modelled upon the pack animal. Again comes the paradox that China, where labour-power is always supposed to have been so abundant, should have been the region where the invention arose. In Europe it may be counted among the humbler machines of the Renaissance, and it undoubtedly aided the industries then developing, but in China it is difficult to point to any disturbing increase in transportation facilities which it assisted. Moreover, the Chinese fitted wheelbarrows with masts and sails, inspiring Milton's famous stanza:

> '. . . the barren plaines
> Of Sericana, where Chineses drive
> With sails and wind their canie waggons light'.

This embodied the misunderstanding that four-wheeled sailing-carriages were involved, an idea which flourished in the ornamentation of many sixteenth-century Western atlases, and directly inspired the Dutch physicist and engineer Simon Stevin to his successful experiments with sailing-carriages upon the sandy beaches of northern Holland. These it was which first demonstrated to Europeans that it was possible for human beings to travel at forty miles or more an hour without perceptible harm, and thus the sailing wheelbarrows of Kiangsi with their loads of porcelain from Ching-tê-chen, though making no particular impression in their own country, struck the imagination of the initiators of that modern science which before long was destined to make aeroplanes fly at four hundred, or rockets (also of Chinese ancestry) at four thousand, miles an hour.

VII

We come now to the second group of technological inventions which deserve attention. A mystery surrounds the origin of rotary milling and the application of water-power to rotary milling, for both these procedures, fundamental in the history of engineering, appear at about the same time in China and the West. One can only assess the fourth to the second century B.C. as the focal period for the former technique, but for the latter we have rather accurate datings. The first water-mill in the West belonged to Mithridates, king of Pontus, about 65 B.C.; the first Chinese water-mills appear about 30 B.C. for working cereal trip-hammer stamp-mills, about A.D. 30 for blowing metallurgical bellows. The difference in date is much too small for direct diffusion in either direction and strongly suggests diffusion in both directions from some intermediate source, but we still have no knowledge of what and where

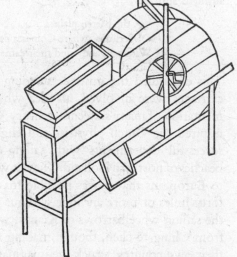

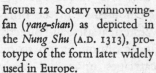

FIGURE 12 Rotary winnowing-fan (*yang-shan*) as depicted in the *Nung Shu* (A.D. 1313), prototype of the form later widely used in Europe.

this was. Whether the most ancient Western water-mills were vertical (i.e. 'Vitruvian', with right-angle gearing) or horizontal ('Norse') we do not know, nor do we know this for China either, save that the shaft which worked the machinery with its lugs must surely have been horizontal (cf. Pl. 8).

Still more fundamental was the invention of the crank or eccentric, and here China's 'legacy' comes in *fortissimo*. For after the rather uncertain types of crank descried by some in ancient Egyptian drilling-tools, the oldest sure examples occur in Han dynasty terracotta models of farmyards with rotary-fan winnowing-machines worked by crank handles (Pl. 9).[1] The oldest European appearance follows after a long period, crank handles for whetstones in the ninth-century Utrecht Psalter.[2] Such an invention is too basic, and at the same time too simple, to leave much trace of its ancient travels, but as a contribution of Chinese technique to that of the Old World as a whole it can hardly be surpassed in importance. In fifteenth-century Europe it generated the crankshaft, a development which in China did not occur, but meanwhile the complete morphology of the reciprocating steam-engine had been perfected there. This requires a little explanation.

Besides the crank, another basic machine had been in use in China since Han times, namely the double-acting piston bellows. There can be little doubt that the very early success of Chinese iron and steel technology had been partly due to this machine, which gave a strong and continuous blast. In addition, the horizontal water-wheel was early in use. These were the components from which one of the most important ancestors of the steam-engine, the hydraulic blower, was constituted. It involved a primary problem of the kinematics of machinery.

For all modern men the most obvious way of converting rotary to longitudinal motion is to use the crank or eccentric, the connecting-rod and the piston-rod—a simple geometrical combination which needs only suitable jointing and the maintenance

---

[1] Critical analysis now dismisses the bailing chain-pump crank handle claimed for the first-century ships of Lake Nemi; and though certain passages in Oribasius and even Archimedes may appear to imply the knowledge and use of it, the philological evidence has not yet been set forth in convincing detail.

[2] The quern with upstanding handle constituted of course a primitive form of crank, but hand-mills of this kind do not antedate the fourth century in Europe, though they are known from the Han period in China (206 B.C.–A.D. 220).

of the piston-rod in a straight line at the end of its back-stroke by means of cross-heads or otherwise. Leonardo da Vinci used this

FIGURE 13 Water-powered blowing-engine (*shui-phai*, or water-powered reciprocator) for blast furnaces and forges as depicted in the *Nung Shu* (A.D. 1313). This is one of the earliest appearances of the conversion of rotary to longitudinal motion by crank, connecting-rod, and piston-rod, hence the inverted predecessor of the reciprocating steam-engine. Motive power is provided by a horizontal water-wheel with a flywheel above it.

system in a saw-mill design towards the end of the fifteenth century in the West, but before his time it cannot be found in Europe. Where one must look for it is at the other end of the Old World, in China, for it appears complete already in 1313 in the agricultural engineering treatise of Wang Chên, who describes it in the form of metallurgical bellows worked by water-power. A horizontal water-wheel drives a flywheel on the same shaft above it, and this in turn rotates by means of a belt-drive a small pulley bearing an eccentric lug; this then works connecting-rod and piston-rod, joined by means of a bell-crank rocking lever (Fig. 13). Thus the entire structure of the reciprocating steam-engine is prefigured in

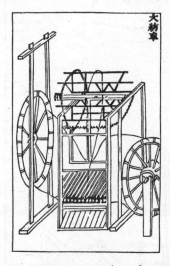 

FIGURE 14 Water-powered textile machinery (spinning machine, *ta fang chhê*) as depicted in the *Nung Shu* (A.D. 1313). The picture on the right illustrates the vertical water-wheel which provides the motive power.

advance, but of course in reverse, for instead of the piston with its rectilinear motion affording the power-source and driving the wheels, the latter, with their rotary motion, drive the piston. Since this engine was in common and widespread use when Wang Chên was writing towards the close of the thirteenth century, it is very unlikely to have originated less than a century or so before-hand, and one may therefore confidently say that the crank, the piston-bellows, and the water-wheel met together to produce the steam-engine's anatomy in the Northern Sung. Water-power was also widely used at this time for driving textile machinery (Fig. 14). I think it can hardly be a coincidence that one of Wang Chên's exact contemporaries was Marco Polo, who was in China while he was writing, or at least meditating, his *Nung Shu*; and since we find very soon afterwards at cities such as Lucca in Italy silk filatures using machinery closely similar to that of China, the presumption is that one or other of the European merchants who travelled East in those days brought back the designs in his saddle-bags.

We have just mentioned silk, and we have also just mentioned

FIGURE 15 Silk-reeling machine (*sao-chhê*) described in the *Tshan Shu* (A.D. 1090).
An early form of flyer, probably the oldest one known, is worked by eccentric
and driving-belt from the main reel shaft, the motive power a treadle crank
'sewing-machine' drive. The single fibres of raw silk are being wound off from
the cocoons in the hot water bath on the left. This illustration, the oldest extant
drawing of the machine, is taken from the first edition of the *Thien Kung Khai
Wu* (Exploitation of the Works of Nature), by Sung Ying-Hsing, A.D. 1637.

driving-belts. There is a more than superficial connexion between them. The domestication of the silkworm and the development of the silk industry had taken place at least as early as the Shang period, in the fourteenth century B.C., and this meant that the Chinese alone were in possession of a textile fibre of extremely long staple. The average length of a single continuous strand of silk amounts to several hundred yards, not at all like a short plant fibre such as flax or cotton, with a staple measurable in inches, which has to be pulled out and spun together to form the yarn. The silk is wound off from the cocoons almost by the mile, and its tensile strength (some 65,000 lb. per sq. in.) far exceeds that of any plant fibre, approaching the level of engineering materials. One can thus begin to understand how it was that the Chinese were so successful in the invention of textile machinery long in advance of other parts of the world. Let us consider, for instance, the *sao-chhê* or silk-winding machine. A text of A.D. 1090, the *Tshan Shu*, by Chhin Kuan, describes it very clearly. The silk is loosened and wound off from the cocoons in the hot water bath; the fibres come up through little guiding rings and are laid down on a great reel. This is worked by a treadle, but its shaft carries also a pulley with a driving-belt which works an eccentric lug on another pulley back and forth, and this in turn operates a ramping-arm to lay down the silk evenly on the reel. We thus have one of the simplest forms of 'flyer'. This is a very important machine from several points of view, partly because it embodies a conversion of rotary to longitudinal motion (though without the piston-rod component) and partly because it is such an early example of the simultaneous combination of motions, one power-source supplying both.[1]

The spinning-wheel is another example, better known, of the driving-belt. We do not yet know whether it took its origin in India, where cotton was indigenous, as is the usual view, or

[1] The example of this usually given (cf. Lynn White, op. cit., p. 119) is the saw-mill of Villard de Honnecourt, c. 1235, where the water-wheel not only powered the excursion of the saw but also assured the feed of the wood. The silk-reeling machine is older by at least two centuries but we do not know when it was first worked by animal- or water-power.

whether it did not rather arise as a quilling-wheel for winding silk on reels in the Chinese culture-area, this apparatus being detectable in literary sources as far back as the Han, and pictured in 1210. The latter is more than possible,[1] for our oldest picture of a spinning-wheel from any civilization is in a Sung painting of 1270, some time (though not perhaps very long) before the first European evidence (Pl. 10). In many Chinese forms the belt runs over three spindles at one time, and the wheel is driven by a treadle with a strange kind of universal joint. Apart from textual references to spinning-wheels near 1300 the first illustrations of driving-belts in Europe come in the fifteenth-century German military engineering manuscripts. There is thus a considerable priority for this essential form of power transmission in Chinese culture, and as usual no good reason for thinking that it was invented later in Europe independently. If then it is true that driving-belts began among Chinese artisans it would not be at all surprising that power-transmitting chain-drives should also be found early among them. And indeed we have already seen precisely this, the chain-drive in the late eleventh-century monumental astronomical clocks of Su Sung (p. 82), probably not new with him but going back at least as early as the similar clockwork of Chang Ssu-Hsün a hundred years before. Endless chains had of course been well known to the Alexandrian mechanicians of the first century B.C., but they were never continuously power-transmitting, and generally more like conveyor belts.

The effects of all these inventions and engineering solutions on the technology of post-Renaissance Europe are self-evident, and the only question the reader will ask is why they did not lead to a similar upsurge of industrialism in China. Here the answer can only be a part of the general observation that Europe had a capitalist revolution (or rather a series of them) and that China did not. Technical novelty alone could no more bring about a

[1] One must not think that the spinning-wheel necessarily arose in connexion with short-staple plant fibres, for the Chinese, who never wasted anything, used it for spinning silk from wild or broken cocoons, and they may well have done this far back into antiquity.

fundamental change in the structure of society than mercantile activity alone or social criticism alone. Something beyond that, some complex of prior conditions, not analysable by us here, would have been necessary in China if the inventions of her brilliant technicians were to have exerted their full effects within her borders. As it was they spread out all over the world as part of her 'legacy'.

VIII

The third group of technological advances centres upon the mastery of iron and steel, but leads us into other fields, some rather unexpected, such as bridge-building and deep drilling. Optimistic American writers of the Jules Verne era were fond of referring proudly to the then modern world as the Iron Age, when the 'iron horse' galloped across the prairies and 'ironclads' began to plough the seas. They would have been surprised to learn that there had been a previous iron age, but not in Europe, rather in medieval China.[1] Until the end of the fourteenth century no European had ever seen a pig of cast iron, yet the mastery of the molten metal had already been achieved in Chinese culture some eighteen centuries before. Of all our paradoxes this is perhaps the most extraordinary, namely that advanced iron-working, so deeply characteristic of developing capitalist industrialism in the West, should have existed for so many centuries within Chinese bureaucratic feudalism without upsetting it.[2]

Iron itself was a relatively late introduction to China, datable in the sixth century B.C. or so, a long time after its twelfth-century B.C. discovery by the Hittites in western Asia Minor. But the remarkable thing is that the Chinese could cast it almost as soon as they knew about it at all. Within two or three centuries the

[1] The reader is referred to the monograph by J. Needham, *The Development of Iron and Steel Technology in Ancient and Mediaeval China* (London, 1958, repr. Heffer, Cambridge, 1964) (Dickinson Lecture).

[2] See a recent interesting paper by R. Hartwell, 'A Revolution in the Chinese Iron and Coal Industries during the Northern Sung Dynasty (960 to 1126)', *Journ. Asian Studies*, 1962, 21, 153.

wrought iron of bloomery furnaces was giving place to cast iron. Among the reasons for this rapid advance we must undoubtedly number the double-acting piston-bellows giving its continuous blast (already mentioned), perhaps also the presence of ores with high phosphorus content, which allows the melting of iron at a temperature about 200° less than otherwise. Besides, one must never forget that the Chinese had been perhaps the greatest bronze-founders of all antiquity, so that much furnace experience, drawn not only from them but from their predecessors the potters, lay at the disposal of the first iron-masters.[1] Furthermore, good refractory clay was available, so that a way of reducing iron ore in crucibles stacked in coal was used at an early date, certainly not later than the fourth century A.D. Archaeological excavations have brought to light quantities of cast-iron tools from the fourth century B.C. onwards, and these are now to be found in many museums in China—hoes, ploughshares, picks, axes, swords, and the like. Remarkable cast-iron moulds have also been found in tombs dating from the late Warring States period, though whether they were used to make cast-iron implements or to cast bronze ones is not yet certain. One or two reliefs of Han date (c. 100 B.C.–A.D. 100) remain which give an idea of the primitive blast-furnaces and bellows used in those ancient times. Our earliest picture of the characteristic small Chinese blast-furnace comes from the *Ao Pho Thu Yung*[2] of 1334, and the best-known one occurs in the *Thien Kung Khai Wu*[3] of 1637. These illustrations show the iron flowing

[1] See N. Bernard, *Bronze-Casting and Bronze Alloys in Ancient China* (Canberra and Tokyo, 1961) (*Monumenta Serica*, monograph series no. 14). Bernard doubts the very existence of a bloomery stage of wrought-iron production before the appearance of iron-casting, and it is true that traces of the earlier phase (naturally assumed in parallelism with development elsewhere) are scarce, nevertheless they occur.

[2] 'The Boiling Down of the Sea', a treatise on the salt industry by Chhen Chhun. There was always a close connexion between salt and iron, partly because in the ancient world of self-sufficient local communities these were the two great commodities which could be produced only at particular places and had to be transported, hence their 'nationalization' in Han times; partly also because the evaporation of the concentrated brine needed large cast-iron pans.

[3] 'The Exploitation of the Works of Nature', a general description of technology and industry by China's Diderot, Sung Ying-Hsing.

out from the blast-furnace and being conducted to the puddling platform for conversion into wrought iron (Fig. 16). Such small blast-furnaces lasted down in many rural districts into the present century, and photographs of them, as also of the crucible method, are available.

Many objects both inside and outside museums testify to the great use made of iron in ancient and medieval China. Han statuettes and vessels beautifully cast in iron have long been known. Then there are the famous funerary kitchen stoves of the Three Kingdoms period (the third century A.D.), the material of which first awakened Western archaeologists to the ancient character of the cast-iron industry in China. Next come many Buddhist statues cast between the fourth and the eighth centuries, often dated, and showing great skill and artistic taste on the part of the craftsmen who made them. The great lion of Tshang-chou, a monument about thrice the height of a man, and one of the largest iron castings in the world, was set up in A.D. 954 by Kuo Jung, one of the Northern Chou emperors, to commemorate his victory over the Liao (Chhi-tan) Tartars. In the Sung a number of cast-iron pagodas were erected, at least two of which still exist complete today (Pl. 11). In the Ming the temples at the top of the sacred mountain Thai-shan were roofed entirely with cast-iron tiles, to withstand the gales that sweep across the summit. All these were peaceful uses. But iron and steel had of course been the basis of the successes of Chinese arms through the ages, whether in the repulsion of the Huns or the Japanese, or the conquests of Sinkiang or Tibet. Protective armour developed more for ships than for men, reaching its apogee in the fleets of armoured vessels under the Korean admiral Yi Sunsin (1585-95). Although it still remains undemonstrated, we can now appreciate better the weight of probability which suggests that the first iron-barrel cannon were Chinese (cf. p. 67 above).

Steel production lagged in no way behind that of iron. In the earliest times steel may have been made by the cementation process as in the ancient West, wrought iron being heated in charcoal to gain the necessary carbon; but as soon as cast iron became abundant

FIGURE 16 Traditional blast-furnace in operation, showing the double-action piston-bellows (here manually operated), the tapping of the cast iron, and its conversion to wrought iron, with use of silica, on the fining platform. From the *Thien Kung Khai Wu* (A.D. 1637; Chhing illustration).

it proved more convenient to oxygenate the product carefully (refining, as it was called in the West), stopping at the stage of steel, with its intermediate carbon content. Then in the sixth century A.D. came the ingenious invention of the co-fusion process (the ancestor of the Siemens-Martin open hearth of today), probably due to a Taoist swordsmith Chhiwu Huai-Wên; here wrought-iron billets and cast-iron chips were heated together in a special furnace. The cast iron melted and bathed the pasty masses of wrought iron so that an interchange of carbon took place, and, upon forging, good eutectoid steel was obtained. So persistent (indeed because so effective) have been some of the ancient

Chinese technical procedures that I myself was able to see in 1958 in Szechuan a closely similar derivative process still successfully at work.

Siderurgical skill had a number of important technical consequences in ancient and medieval China. The availability of excellent wrought-iron chains suggested their use for a fundamental improvement of the suspension bridges of bamboo cables by means of which since ancient times many of the rivers flowing through ravines in western China had been crossed by the main lines of communication. The textual and archaeological evidence available indicates that it was in the Sui Dynasty (A.D. 589–618) that iron-chain suspension bridges first spanned the necessary 200–300 ft. gaps, but there can be no doubt whatever that these bridges were common during the Sung, Yuan, and Ming periods. The first proposal in Europe was made by the engineering bishop Faustus Verantius about 1595, but no successful bridge was built until the middle of the eighteenth century (1741). It is exceedingly likely that Verantius had heard of the Chinese bridges from the early Portuguese travellers, and it is certain that Fischer von Erlach who described and recommended them in 1725 drew from Chinese sources.

Iron entered into bridge construction in yet another manner, also (and still more certainly) attested for the Sui period. This was the time of activity of a very brilliant engineer Li Chhun, who first threw across a river valley a segmental arch bridge with relieving arches in the spandrels. This superb structure, which still exists at Chao-hsien, and has recently been thoroughly repaired, resembles nothing so much as the bold railway bridges in stone or reinforced concrete constructed since the seventies of the last century; and with its group of similar structures in North China must surely have exerted an influence on the builders of the first segmental arch bridges of Europe, the Ponte Vecchio at Florence (1345) and its successors. Li Chhun's audacious design was assisted by the use of iron clamps between the stones of the twenty-five parallel arches of which the bridge vault was composed (Pl. 12).

The connexion between iron and salt has already been noted;

large cast-iron pans were needed for the evaporation of the brine. But there was another more curious link. At an early time it was found that Szechuan province, a couple of thousand miles away from the sea, possessed great stores of natural brine and natural gas in pockets far below the surface of the red-earth basin. The exploitation of this began at least from the beginning of the Han period (second century B.C.), as we know both from textual and archaeological evidence (moulded bricks); and a limiting factor here which soon permitted the drilling of deep bore-holes as far down as 2,000 ft. was the availability of good steel for the bits and drilling tools. The method of drilling has often been described; a group of men jump on and off a beam to give an up-and-down movement, while at the same time the drilling cable is

FIGURE 17 The technique of deep drilling practised in Szechuan province from Han times onward for obtaining brine and natural gas. Here the just lengthened bamboo spear or haft of the drilling tool is being cautiously lowered during drilling operations. (*Thien Kung Khai Wu*, A.D. 1637; Chhing illustration).

rotated by another. When the bore-hole is completed, a process that may take several years, a long bamboo tube with a valve is sent down to act as a bucket and bring up the brine.[1] The natural gas, collected from other bore-holes, is used for the evaporation. There can be little doubt that knowledge of these methods spread from China to inspire the drilling of the first artesian well near Lillers in 1126, and there is none at all that the first petroleum wells in America's south-western states were drilled by the ancient Chinese method, there known as 'kicking her down'.

Putting these facts together, we have only to add the certainty that cast iron began to flow from the first blast-furnaces in Europe in the neighbourhood of 1380, mostly in Flanders and the Rhineland, and we know also that one of the great urges for the adoption of this new technique was the desire to cast iron cannon. In view of the long prior history of iron, and especially of iron-casting, in China, I am not disposed to entertain any belief in an independent invention in Europe; at the same time we still know little or nothing of the intermediaries through which the knowledge and experience came. One suspects the Turks, with whom some of the earliest European iron-masters had studied, as also the Persians, but nothing very definite can be said. The other outstanding problem of course is how it was possible for Chinese administration to be so stable in the presence of a metal which had such earth-shaking effects in Europe. To begin with we must remember that Chinese iron-working preceded by some time the first unification of the land into a single empire in the third century B.C.; the state of Chhi had waxed wealthy on iron (as well as salt), but the state of Chhin, which conquered all the other states, probably had a metallurgical policy more strictly directed to its military use. Iron has been called by ancient historians of the West 'the democratic metal',[2] for its widely distributed ores could be acquired and used by city-states and peasant barbarians alike against the older unified monarchies. Since iron was much superior to bronze for weapons this was a grave matter. But in East Asia

[1] Note once again the great value for Chinese technology of the natural tube constituted by the bamboo.
[2] Cf. V. Gordon Childe, *What Happened in History* (London, 1942; American ed. 1946), p. 176.

the whole city-state conception was quite foreign to Chinese culture, and the unified Chhin empire simply took over the bureaucratic anti-aristocratic anti-mercantile ethos which had already grown up within the so-called feudal states. At the same time the barbarian tribes both within and without the empire were kept under control, until the fourth century. The 'nationalization' of iron (as well as salt, fermented beverages, etc.) which had been discussed in Chhi[1] became a reality under the stable dynasty of the Han, and about 120 B.C. all iron production was carried on in forty-nine government factories scattered throughout the empire.[2] Though this was freed in later dynasties, and although doubtless in times of partition certain states benefited from iron and steel supplies which others did not have, individual iron-masters were no more in a position than any other merchant-entrepreneurs to challenge the overwhelmingly bureaucratic domination of the scholar-gentry. And this was just as true after the invention of gunpowder as it had been before. In a word, like the legendary ostrich, Chinese culture could digest cast iron and remain unperturbed thereby; Europe's indigestion amounted to a metamorphosis.

IX

The last group of technical innovations to be considered is connected with the sea. All too unjustly have the Chinese been dubbed a non-maritime people. Their ingenuity manifested itself

[1] Cf. Than Po-Fu, Wên Kung-Wên, Hsiao Kung-Chüan & L. Maverick, *Economic Dialogues in Ancient China; Selections from the Kuan Tzu Book* (New Haven, 1954).

[2] Everyone should read the *Yen Thieh Lun* (Discourses on Salt and Iron), written by Huan Khuan in *c.* 80 A.D., for a partial translation by E. M. Gale has long been available (Leiden, 1931). Additional chapters were translated by E. M. Gale, P. A. Boodberg & T. C. Lin in *Journ. Roy. Asiat. Soc.* (*North China Branch*), 1934, 65, 73. It is the almost verbatim report of a discussion between feudal-minded Confucian scholars and bureaucratic officials concerning the 'nationalized' industries. Some of the problems raised, such as that of standardization of spare parts, have an extraordinarily modern ring. For a further commentary see Chang Chun-Ming in *Chinese Social and Political Review*, 1934, 18, 1.

in nautical matters just as much as elsewhere, the number of their vessels on the inland waters was found by medieval and Renaissance Western merchants and missionaries almost beyond belief, and their sea-going navy was assuredly the greatest in the world between 1100 and 1450.

It all began with the bamboo, the buoyancy of which was soon found useful for the construction of vessels. The bamboo sailing-raft, still characteristic of the South China and Indo-Chinese coasts and Taiwan, is very ancient in date; indeed it has been important in fishing and trading for nearly three millennia. According to a generally accepted opinion, all ship-building in the Western world derived from the dug-out canoe, strakes being built up on each side of it to obtain the wooden ship (whether carvel or clinker) with keel, stem-post, and stern-post. None of these parts exists in the typical Chinese ship (*chhuan*, hence the word 'junk'), which seems to have grown up rather in the form of a rectangular box based on the original bamboo raft. The box shape with transom stem and stern is profoundly characteristic of the junk. Hence arose the segmental construction, the hold being divided by transverse bulkheads. These watertight compartments were, we know, adopted in European shipping in the early nineteenth century with full consciousness of the prior Chinese practice. From the square-ended transom stern another remarkable consequence followed. Although there is no stern-post, the aftermost member, or nearly aftermost member, of the bulkhead series, being vertical, permitted the attachment of a 'stern-post' rudder. Some years ago, my collaborators and I built up an elaborate argument from textual sources showing that the stern-post rudder originated in the Chinese culture-area. This conclusion was then strikingly confirmed by the excavation of model ships in terracotta from tombs in Canton belonging to the first century B.C. and the first century A.D., as we saw to our delight when visiting the museum there in 1958 (Pls. 13, 14). The stern-post rudder then first appears in Europe about 1180, a time almost exactly identical with the appearance and adoption of the magnetic compass there. Of the obvious importance of the latter for the discovery of the route round Africa and the way to the New World we have already spoken in connexion with Francis Bacon's aphorism (pp. 62, 75), but nautical historians agree that the former can have been no less important.

So far we have spoken of structure and guidance, but propulsion is equally important. It is worth emphasizing that all Chinese history knows nothing of the multi-oared slave-manned galleys of the Mediterranean, so prominent in Renaissance as well as Greek sea history. Though it is true that the hauling or 'tracking' of ships up the great rivers and through the rapids was done in all ages by gangs of pullers (free men, nevertheless, in so far as anyone in feudal-bureaucratic clan-family society could be called free), by and large the universal method of propulsion, from the Tung-thing lake to Zanzibar, was sail. Moreover, apart from the sprit-sail, which seems to have been used occasionally in the Hellenistic world, Chinese waters saw the first fore-and-aft sails in the third century, as we know from contemporary textual descriptions. The Chinese were the great proponents of the lug-sail, and great use was made of bamboo up aloft, for the lugs took the form of flat, aerodynamically efficient mat-and-batten sails. The models of the five-masted Shantung trader of fifty years ago in the Science Museum at South Kensington and the National Maritime Museum at Greenwich give one a good idea of what the ships carrying a thousand men or more in the expeditions of the great admiral Chêng Ho in the early fifteenth century must have been like (cf. Fig. 18). Those were the days when the Chinese navy was visiting everywhere from Kamchatka to Madagascar. The physics and mathematics of sails are still imperfectly understood, less well perhaps than those of aeroplane wings, but it is certain that medieval junks could sail well to windward, as the square-sailed cogs of Hanseatic or Catalan Europe could not. Modifications of the Chinese mat-and-batten system have been adopted on many modern racing yachts, notably Hasler's *Jester*, famous for a single-handed Atlantic crossing in 1961. And while speaking of aeroplane wings it may be well to recall that the notable invention of anti-stalling wing-slots is said to have been inspired by the fenestrated rudder of the Chinese junk. For Chinese sea-captains and river-junk masters had long ago found advantage not only in having their rudders balanced (part being forward of the post) but in having them perforated with holes as well.

One last invention must be referred to, that of the paddle-wheel boat. Descriptions of treadmill-operated paddle-wheel boats begin to appear in Chinese literature in the fifth and sixth centuries A.D., and their construction and habitual use for naval combat on lakes and rivers is quite indubitable in the eighth, when a Thang prince, Li Kao, built and commanded a fleet of them. In the twelfth century, when the Chinese navy began to develop prodigiously after the capture of Kaifêng by the Chin Tartars and the retirement of the Sung Dynasty south of the

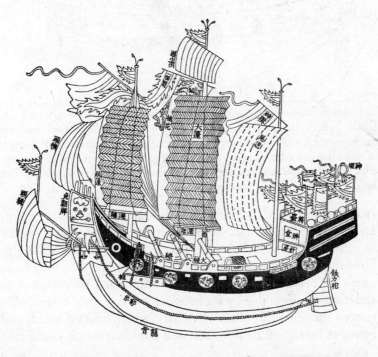

FIGURE 18   An eighteenth-century drawing of a three-masted sea-going junk, from the *Liu-chhiu Kuo Chih Lüeh*, A.D. 1757.

Yangtze, the paddle-wheel warship came into its own.[1] Since iron wheels were difficult to make, and since no substantial power-source was available, the number of wheels was multiplied, and in a campaign against one of the perennial rural rebellions, about 1130 (under Yang Yao), government battleships with as many as

FIGURE 19 Sketch reconstructing the probable appearance of one of the twenty-three-wheel treadmill-operated naval paddleboats used in the Southern Sung, as built by Kao Hsüan and others, A.D. 1130. The casing is shown removed from the stern-wheel and from the six forward wheels on the port side. The number of trebuchets and other weapons is only suggested. Sails were mainly for emergency use, and masts carried crow's-nests. Probable dimensions: length 100 feet, beam 15-20 feet (original drawing).

twenty-three wheels (eleven on each beam and one at the stern) were constructed by a naval architect named Kao Hsüan. As he was captured by the rebels soon afterwards and proved willing to build ships for them too, the campaign dragged on for a long time, much use being made of gunpowder bombs and poison-smokes; and it was finally ended by the famous loyalist general Yo Fei, who enticed the rebel ships into an estuary where floating weeds and branches entangled their wheels and boarding action could take

[1] See the interesting paper by Lo Jung-Pang entitled 'China's Paddle-Wheel Boats: the Mechanised Craft Used in the Opium Wars, and their Historical Background', *Tsinghua Journ. Chinese Studies*, N.S., 1960, **2**, 189.

place. Paddle-wheel warships continued to be of much importance, however, guarding the Yangtze for a century and a half so well that no further penetration of the Chin Tartars to the south occurred, river-crossings being impossible. Afterwards, with the general conquest of the country by the Mongols, the age of the paddle-wheel ship declined, since they were more interested in sea-fighting (cf. the attempted conquests of Java and Japan), where, without a power-source such as steam, the paddle-wheel was unsuitable. Again we have no knowledge to what extent these traditions influenced the first trials in Europe, which took place in 1543 at Barcelona. They were certainly handed down in China, for during the Opium Wars (1839–42) a considerable number of treadmill-operated paddle-wheel war-junks were sent against the British ships and gave a good account of themselves, though the cause was hopeless. With their usual complacency the Westerners supposed that these ships had been built in imitation of their own paddle-steamers, but study of the contemporary Chinese literature shows that this was not so at all. The whole story has the further interesting feature that there had been a proposal in fourth-century Byzantium for a paddle-wheel ship to be worked by ox whims, but there is no evidence of its ever having been constructed. As the manuscript was only discovered at the Renaissance, it cannot have influenced the Chinese ship-builders. How far it influenced the Barcelona experiment remains uncertain, for there had been proposals for paddle-wheel boats in the German engineering manuscripts of the fifteenth century, and these may have been re-inventions inspired by the omnipresent vertical water-mills. In any case, there is no doubt at all that though the first suggestion was Byzantine the first practical realization was Chinese.

X

This brings us to the term of our survey of China's 'legacy'. Before dwelling upon the principal paradoxes which emerge from it, we must take notice of a curious, and perhaps significant, fact; namely that it is possible to distinguish, at least in the technological field, arrivals of innovations from Asia, mainly China, in particular collocations which I have come to call 'clusters'. Thus for example between the fourth and the sixth centuries A.D. one finds the arrival of the drawloom and the breast-strap harness. By the eighth century the foot-stirrup is exerting extraordinary effects, and soon afterwards the Cardan suspension appears. By the beginning of the tenth century equine collar harness has come, together with the simple trebuchet in the field of artillery. The eleventh century sees the spread of the Indian numerals, place-value, and the zero sign. Towards the end of the twelfth century come in a cluster the magnetic compass, the stern-post rudder, paper-making, and the idea of the windmill, with the wheelbarrow and the counter-weighted trebuchet quickly following; this was the time of the Toledan Tables. Towards the end of the thirteenth century and the beginning of the fourteenth there appear in another cluster gunpowder, silk machinery, the mechanical clock, and the segmental arch bridge; this was the time of the Alfonsine Tables. Rather later, but still forming part of this second inflow, we find the blast-furnace for cast iron, and block-printing, soon followed by movable-type printing. During the fifteenth century the standard method of interconversion of rotary and rectilinear motion establishes itself in Europe, and other East Asian engineering motifs appear, such as the spit vane-wheel, the helicopter top, the horizontal windmill, the ball-and-chain flywheel, and lock-gates in canals. The sixteenth century brings the kite, the equatorial mounting and equatorial co-ordinates, the doctrine of infinite empty space, the iron-chain suspension bridge, the sailing-carriage, a new emphasis on sphygmology in medical diagnosis, and equal temperament in musical acoustics. In the eighteenth century a rearguard is formed by variolation (the predecessor of vaccination), porcelain technology, the rotary-fan winnowing-machine, watertight compartments at sea, and some other late introductions such as medical gymnastics and ultimately the system of civil service examinations.

This list of transmissions, though very incomplete, throws into some relief the historical fluctuations in the reception of East

Asian discoveries and inventions by Europe. Though it is generally impossible to trace the course of a 'blue-print', or a stimulating idea, still less the mere conviction that a problem had already been successfully solved somehow, the general circumstances which facilitated flow at particular times present themselves as obvious— for the twelfth-century cluster the Crusades, the Qara-Khitai kingdom in Sinkiang, etc.; for the fourteenth-century cluster the Pax Mongolica; for the fifteenth century the Tartar slaves in Europe; for the sixteenth century onwards the Portuguese travellers and the Jesuit mission, and so on. The earlier periods of transmission are more obscure, and further research will be needed to elucidate them, but an overall picture of the world's indebtedness to East Asia, especially China, is emerging very clearly.

The first of the paradoxes with which I wish to conclude is that according to a common belief there was never any science or technology in China at all. It may seem passing strange in the light of all that has gone before that this should ever have been believed, yet such was the impression of my sinological elders when I began the investigation of these matters, and it has been enshrined in one form or another in many well-known statements. Repeated as it was by generations of superficial observers of Chinese everyday life who knew nothing of the literature, the Chinese ended by believing it themselves. An essay by the great Chinese philosopher, Fêng Yu-Lan, written more than forty years ago, bore the title 'Why China has no Science'.[1] In this he said:

'I shall venture to draw the conclusion that China has had no science, because according to her own standard of values she has not needed any. . . . The Chinese philosophers had no need of scientific certainty because it was themselves that they wished to know; so in the same way they had no need of the power of science, because it was themselves that they wished to conquer. To them, the content of wisdom was not intellectual knowledge, and its function was not to increase external goods.'

There was of course something in this, but only something, and the standpoint may have been influenced by a feeling that what

[1] *International Journ. Ethics*, 1922, **32** (no. 3), reprinted (in English) in his collected essays, *Chung-kuo Chê-hsüeh Shih Pu* (Contributions to the History of Chinese Philosophy), Shanghai, 1936.

China did not seem to have happened to have was not very much worth having anyway.[1] The converse of Fêng Yu-Lan's youthful pessimism is to be seen in the equally unjustified optimism of Arnold Toynbee:[2]

'However far it may or may not be possible to trace back our Western mechanical trend towards the origins of our Western history, there is no doubt that a mechanical penchant is as characteristic of the Western civilization as an aesthetic penchant was of the Hellenic, or a religious penchant was of the Indic and the Hindu.'[3]

Today it has become quite evident that no people has had a monopoly of philosophical mysticism, scientific thought, or technological ability. The Chinese were not so uninterested in external nature as Fêng Yu-Lan averred, and the Europeans were by no means so ingenious and inventive as Toynbee claimed. The paradox may have arisen partly from a confusion in the meaning of the word 'science'. If one defines science as modern science only, then it is true that it originated only in Western Europe in the sixteenth and seventeenth centuries in the late Renaissance, the life of Galileo marking the turning-point. But that is not the same thing as science as a whole, for in all parts of the world ancient and medieval peoples had been laying the foundations for the great building that was to arise. When we say that modern science

[1] A related version of this theme, not infrequently stated, is that in the traditional Asian cultures 'it was felt easier for man to adapt himself to Nature than to adapt Nature to himself'. I quote from Alan Watts, Nature, Man and Woman (London, 1958), p. 52, a brilliant and, in other respects, a very perceptive book. This particular thesis is falsified by twenty centuries of Chinese scientific and technological history.

[2] A. J. Toynbee, A Study of History (6 vols.; London, 1935–9), vol. iii, p. 386.

[3] A related version of this theme has become well known as the thesis of F. S. C. Northrop (see especially his The Meeting of East and West; an Enquiry concerning Human Understanding, New York, 1946). According to him the Greeks developed the way of knowing Nature by rational postulation and scientific hypothesis, while the Chinese throughout their history approached Nature only by direct observation, empathy, and aesthetic intuition. This is just as untenable as the rest. A cruder, more racialist, formulation of the same ideas occurs in the stimulating, but wholly unreliable, book by L. Abegg, The Mind of East Asia (London, 1952), cf. pp. 233 ff., 294 ff., etc.

developed in Western Europe in the time of Galileo, we mean most of all, I think, that there alone there developed the fundamental principle of the application of mathematized hypotheses to Nature, the use of mathematics in putting questions, in a phrase, the combination of mathematics with experiment. But if we agree that at the Renaissance the method of discovery was itself discovered, we must remember that centuries of effort had preceded the break-through. Why this happened in Europe alone remains a subject for sociological investigation. We need not here prejudge what such investigations will reveal, but it is already obvious enough that Europe alone underwent the combined transfigurations of the Renaissance, the scientific revolution, the Reformation, and the rise of capitalism. These were the most extraordinary of all the phenomena of Western instability before socialist society and the atomic age.

But here comes the second paradox. From all that has been said it is clear that between the fifth century B.C. and the fifteenth century A.D. Chinese bureaucratic feudalism was much more effective in the useful application of natural knowledge than the slave-owning classical cultures or the serf-based military aristocratic feudal system of Europe. The standard of life was often higher in China; it is well known that Marco Polo thought Hangchow a paradise. If there was on the whole less theory there was certainly more practice. If the scholar-gentry systematically suppressed the occasional sprouts of mercantile capital, it was seemingly not in their interests to suppress innovations which might be put to use in improving the production of the counties or provinces in their charge. If China had an apparently limitless reservoir of labour-power it remains a fact that we have so far met with no single case of the refusal of an invention due to explicit fear of technological unemployment. Indeed the bureaucratic ethos seems to have helped applied science in many ways. One could instance the use of the Han seismograph to signalize and locate calamities before news of them reached the capital, the erection of a network of rain-gauge and snow-gauge stations in the Sung, or the extraordinary expeditions undertaken in the

Thang to measure a meridian arc from Indo-China to Mongolia over 1,500 miles long,[1] and to map the stars from Java to within 20° of the south celestial pole. The *li* was keyed to a celestial-terrestrial standard a hundred years before the kilometre. Let us not despise then the mandarins of the celestial empire.

And so we come at last to the paradox of paradoxes—'stagnant' China the donator of so many discoveries and inventions that acted like time-bombs in the social structure of the West. The cliché of stagnation, born of Western misunderstanding, was never truly applicable; China's slow and steady progress was overtaken by the exponential growth of modern science, with all its consequences, after the Renaissance. To the Chinese, could they have known of her metamorphoses, Europe would have seemed a civilization in perpetual upheaval;[2] to Europeans, when they came to know her, China seemed to have been always the same. Perhaps the stereotypic inanity among Western commonplaces is the belief that although the Chinese invented gunpowder they were so foolish— or so wise—that they used it only for fireworks, leaving its full powers to be exploited by the West alone.[3] We may not wish to deny to the West a certain penchant, alas, for *Büchsenmeisterei*, but the idea behind the commonplace is of course that without the West nothing grand or creative would have been done with such inventions. The Chinese made sure that their tombs faced due south, but Columbus discovered America; the Chinese planned the steam-engine's anatomy, but Watt applied steam to the piston; the Chinese used the rotary fan but only for cooling palaces,[4] the Chinese understood selection but confined it to the breeding of

---

[1] Cf. the detailed account by A. Beer, Lu Gwei-Djen, J. Needham, E. Pulleyblank, and G. I. Thompson, 'An Eighth-Century Meridian Line . . .', *Vistas in Astronomy,* 1961, **4**, 3.

[2] It was not that China knew no upheavals of civil strife, dynastic change, and foreign invasion, quite the contrary, but the basic forms of her social life remained relatively constant.

[3] This appears, of course, in the book by L. Abegg previously quoted, as but one exhibit in a museum of similar statements (p. 235).

[4] This they did, but they also used the encased and crank-operated rotary winnowing-fan some eighteen centuries before Europeans got hold of it.

fancy goldfish.[1] All such fancied antitheses are demonstrably historically false. The inventions and discoveries of the Chinese were mostly put to great and widespread use, but under the control of a society which had relatively very stable standards.

There can be no doubt that there was a certain spontaneous homoeostasis about Chinese society and that Europe had a built-in quality of instability. When Tennyson wrote his famous lines about 'the ringing grooves of change' and 'better fifty years of Europe than a cycle of Cathay',[2] he felt impelled to believe that violent technical innovation must always be advantageous; today we might not feel quite so sure. He saw effects only, ignoring causes, and moreover in his time physiologists had not yet come to understand the constancy of the internal environment,[3] nor engineers to build self-regulating machines.[4] Cathay had been self-regulating, like a living organism in slowly changing equilibrium, or a thermostat—indeed the concepts of cybernetics could well be applied to a civilization that had held a steady course through every weather, as if equipped with an automatic pilot, a set of feedback mechanisms, restoring the *status quo* after

[1] This they did, but far more important was the process of improvement of rice and other staple crops, which went on for centuries with very conscious imperial supervision and encouragement.

[2] Is it possible that Tennyson knew there was in fact a real Chinese cycle of sixty years? More probably he was thinking of the *kalpa* and the *mahākalpa*. But the mistake, if such it was, illuminates strangely the theme of this paper.

[3] We now know that living organisms maintain a constancy of internal conditions, including the composition of their body-fluids, automatically regulating temperature, pressure, acidity, blood-sugar, etc.; and that they do this the more effectively the higher in the evolutionary scale they are. There is always great danger in applying biological analogies to social phenomena, as I have often in the past pointed out (cf. *Time, the Refreshing River* (London, 1943), pp. 114 ff., 160 ff.; *History is on Our Side* (London, 1946), pp. 192 ff.). Nevertheless in the present case it seems to me that the replacement of the false and meaningless concept of 'stagnation' by the precise and applicable idea of slowly changing 'homoeostasis' brings a definite increase in the clarity of our thought about traditional Chinese culture.

[4] This is not quite true historically, for the fantail gear of the windmill and the governor-balls of the steam-engine had both been long in use by Tennyson's time. But their philosophical significance was hardly noticed, and self-regulating machinery did not impose itself on general thinking until the era of electrical communication-, as opposed to power-, technology.

all perturbations, even those produced by fundamental discoveries and inventions. Struck off continually like sparks from a whirling grindstone, they ignited the tinder of the West while the stone continued on its bearings unshaken and unconsumed. In the light of this, how profoundly symbolic it was that the ancestor of all cybernetic machines, the south-pointing carriage, should have been a Chinese device.[1]

There was no special superiority about the relatively 'steady state' of Chinese society, resembling as it did in many ways ancient Egypt, that age-old continuum which amazed the youthful, changeable Greeks. The constancy of the internal environment is only one function of living organisms, necessary but not as complex for example as the higher activities of the central nervous system. Metamorphosis is also a perfectly physiological process, and in some living things it can go as far as the complete dissolution and re-formation of all the tissues of the body. Perhaps civilizations, like different kinds of living beings, have developmental periods very different in length, and when they metamorphose do so in varying degrees.

There is no special mystery about the relatively 'steady state' of Chinese society either.[2] Social analysis will assuredly point to the nature of the agriculture, the early necessity of massive hydraulic engineering works, the centralization of government, the principle of the non-hereditary civil service, etc., etc. But that it was radically different from the patterns of the West is quite unquestionable.

To what then was the instability of Europe due? Some have referred it to the aspirations of the never-satisfied Faustian soul.

[1] This invention, which took place in China in the third century if not a little earlier, has been mentioned in passing, above, p. 60. A carriage bore a figure which pointed to the south, and continued to do so in whatever direction the carriage was made to move. It is certain that this was accomplished mechanically, probably by a simple form of differential gear, and it is very likely that the inventor was Ma Chün.

[2] In this paper I have perhaps overemphasized the continuity and unity of Chinese society. Byzantine society, as Gibbon painted it, seemed similarly 'monolithic', but modern research has revealed great variations of its structure during its different periods. China too will show changes of finer structural detail 'under higher power', as microscopists say, but some simplification was here unavoidable.

I would prefer to think in terms of the geography of what was in effect an archipelago, the perennial tradition of independent city-states based on maritime commerce and jostling military aristocrats ruling small areas of land, the exceptional poverty of Europe in the precious metals, the continual desire of Western peoples for commodities which they themselves could not produce (one thinks especially of silk, cotton, spices, tea, porcelain, and lacquer), and the inherently divisive tendencies of alphabetical script, which permitted the growth of numerous warring nations with centrifugal dialects or barbarian languages. By contrast China was a coherent agrarian land-mass, a unified empire since the third century B.C. with an administrative tradition unmatched elsewhere till modern times, endowed with vast riches both mineral, vegetable, and animal, and cemented into one by an infrangible system of ideographic script admirably adapted to her fundamentally monosyllabic language. Europe, a culture of rovers, was always uneasy within her boundaries, nervously sending out probes in all directions to see what could be got—Alexander to Bactria, the Vikings to Vineland, Portugal to the Indian Ocean. The greater population of China was self-sufficient, needing little or nothing from outside until the nineteenth century (hence the Hon. Company's opium policy), and generally content with only occasional exploration, essentially incurious about those far parts of the world which had not received the teachings of the Sage.[1] Europeans suffered from a schizophrenia of the soul, oscillating for ever unhappily between the heavenly host on one side and the 'atoms and the void' on the other; while the Chinese, wise before their time, worked out an organic theory of the universe which included Nature and man, church and state, and all things past, present, and to come.[2] It may well be that here, at this

[1] This must not be exaggerated. It is well to remember that Europe was discovered by China and not the reverse—when Chang Chhien travelled through Central Asia from 138 to 126 B.C. and came upon Greek Bactria. Besides, there were periods of great Chinese enterprise in voyaging, for example the exploits of the Ming navy under Chêng Ho in the early fifteenth century (p. 110).

[2] Hence the rather naïve conception of Laws of Nature enacted by a supernatural law-giver did not develop among them. But there is no doubt that this idea had great heuristic value in the initial phases of modern science.

point of tension, lies some of the secret of the specific European creativeness when the time was ripe. In any case it was not until the flood-tide of that modern science and industry so generated washed away her sea-walls that China experienced the necessity of entering the world *oikoumene* which these great forces were forming. And so her legacy joined with those of all the other cultures in a process which is palpably bringing into being the world co-operative commonwealth.

3

ON SCIENCE AND SOCIAL CHANGE

Written at Wa-yao, Yunnan, when cut off by landslides on the Burma Road, Sept. 1944. Intended for
a collective work edited by C. C. Lienau.

First published in *Science and Society*, 1946, **10**, 225.

I

Those whose work has lain in China around the year 1944 have
been exceptionally isolated. A large part of my work, indeed,
consists in trying to break the blockade which cuts off Chinese
scientists and technologists from their colleagues in the other
United Nations. At the moment of writing, furthermore, I find
myself in a vacuum within a vacuum, cut off from all communication with the outside world by landslides and rockfalls, without
telegraph or telephone, awaiting the clearing of the road. Such
conditions, however, are propitious for writing some paragraphs
of the kind required for a volume of essays on social organization,
and in fact had it not been for a fortuitous few days of enforced
idleness, the pressing work of wartime duties would not have
permitted the writing of this contribution.

In books such as *Time, the Refreshing River* and *History is on
Our Side*, I have tried to outline the characteristics of the various
levels of integration which we find in the world around us,
and the profound effects which a proper appreciation of them has
on one's world outlook and actions in society. In doing so one's
feelings are rather mixed, for the conception of levels seems so
painfully obvious, and yet at the same time so many of the philosophies and world outlooks which intelligent people entertain

123

seem to be based on a complete disregard of the world's development, past and to come. It need hardly be said that these ideas did not originate with the present author. They are to be found in the dialectical materialism of Marx and Engels, in the organic mechanism of Whitehead, the axiomatic biology of Woodger, the evolutionary naturalism of Sellars, the emergent evolutionism of Lloyd Morgan and Samuel Alexander, the holism of Smuts. Each of these philosophers emphasized different aspects, but some deserve greater credit than others in that they were bold enough to draw conclusions applicable to our personal actions in the real world we live in.

However, in order to make clear what lies behind the point of view of this essay, it is necessary once more, at risk of unwelcome repetition, to outline what I mean by integrative levels. The levels occur in the form both of *envelopes* and *successions*. Spatially, the smaller organisms are contained in the larger. The physical particles are in the atoms, the atoms are in the molecules, the molecules are in the colloidal aggregates, the latter are in the living cells, the cells are within the organs, and these again within the living bodies, and finally the bodies are within the social aggregates, of which there are aggregates of aggregates, organisms of organisms, up to the highest of which we can conceive. All kinds of organisms find their place in such a scheme—insect societies, ecological associations, groups of cells explanted *in vitro*, determined and undetermined transplants, polymer molecules, liquid crystals, bombarded and disintegrating atoms, bombed yet still functioning cities, traffic in blood-vessels or arterial highways, sessions of scientific societies, visions of the World Co-operative Commonwealth to come.

According to Whitehead's immortal dictum, physics is the study of the simpler organisms and biology is the study of the more complicated ones. But the other aspect of this view is the temporal one. There has been a succession in time of these various levels. There were physical particles before there were atoms, simple atoms before there were large unstable ones, molecules before there were living cells and protoplasm, living particles or cells in isolation before there were metazoa and metaphyta, and primitive plants and animals before there were the most highly complex and active ones.

The only obvious guiding thread running through this series of stages of order, is a rise, in spite of all setbacks, in the level of organization. As a biochemist, the writer has tried to define it especially in the biological field. One may perhaps say, as in *Time, the Refreshing River*,[1] that as we rise in the evolutionary scale from the viruses and protozoa to the social primates there is: (1) a rise in the number of parts and envelopes of the organism and the complexity of their morphological forms and geometrical relations; (2) a rise in the effectiveness of the control of their functions by the organism as a whole; (3) a rise in the degree of independence of the organism from its environment, involving diversification and extension of range of the organism's activities; (4) a rise in the effectiveness with which the individual organism carries out its purposes of survival and reproduction, including the power of moulding its environment. There is a sense, of course, in which an amoeba is as organized as a man in that it carries out all the functions of assimilation, reproduction, metabolism, etc.; but the difference lies in the variety of conditions under which it can do so, and the kind of limitations on the type of life which it can lead. There is also a sense in which *all* those species of plants and animals which have succeeded in persisting through evolutionary change are equally successful. But this is not the only criterion of success. Merely to persist is certainly the *sine qua non*, but we have also to consider under what variety of changed circumstances this persistence can occur, and also what the organism does with its persistence. Limpet's vegetative life is clearly simpler than fox's indefatigable prowling or man's construction of a Golden Gate Bridge.

A. J. Lotka, in an article published in this journal, commenting on one of the essays in *Time, the Refreshing River*, suggests[2] that the direction of organic evolution is such as to make the flux of energy through the system of organic nature a maximum. Unless

[1] P. 211.   [2] *Science & Society*, 1944, **8**, 168.

I misunderstand him, this idea is akin to that of 'life more abundant' expressed gropingly in the previous paragraph. 'This surmise,' he says, 'is based upon the fact that the predominantly anabolic plants tend, in spreading over the globe, to increase the energy absorbed by organic nature per unit of time, while the advent and extension of animals, with their pronounced catabolic activities, tend to increase the energy dissipated per unit of time. It is as if a reservoir were put in the path of an abundant stream, and both the inlet and the outlet of the reservoir were enlarged, thus increasing the flow through it.' By some such method as this it might be possible to find a quantitative formulation of increasing organizational level. The law of organic evolution must be expressed in the form that some function of the evolving system as a whole is proceeding towards a maximum. When we are able to read the book in its entirety to which this essay is a contribution, we may be better able to judge of the feasibility of Lotka's suggestion.

It is important to note, however, that Lotka's formulation is that of a thermodynamic statistician while mine was that of a chemical biologist. Neither of them takes account of that cosmic, pre-biological, increase of organization involving physical particles, atoms, and molecules, to which reference was made above when the envelopes and successions were described, and which L. J. Henderson had so much in mind in his *Fitness of the Environment* and his *Order of Nature*. Something more fundamentally geometrical, topological, or morphological than Lotka's energetical formulation would seem to be necessary to cover the whole range.

Perhaps the other most important corollary of the conception of integrative levels is the entire liquidation of all controversies of the 'vitalism-mechanism' type. It is useless to argue whether a science dealing with one integrative level can be 'reduced' to that dealing with another. Science at each level has to work with the concepts and tools and laws appropriate to that level. In genetics and embryology, for instance, we know of many regularities which cannot be affected by anything that biochemistry may discover. But on the other hand they can only attain their full meaning and significance in the light of what biochemistry will discover. Similarly, chemistry is independent of, but only fully meaningful in the light of, physics; and sociology in that of biology. The laws appropriate to the higher or coarser levels cannot be elucidated by observing the lower or finer levels. The higher or coarser levels cannot be explained in terms of the lower or finer ones, still less *vice versa*; but the universe remains quite enigmatic until we succeed, as from time to time we do, in finding the relations between various behaviours at two adjacent levels. This whole subject is very important, as the neglect of it has led to notable sociological heresies, such as the Nazi-fascist determination to apply purely biological standards to human societies, or the use of physico-chemical concepts for such high organisms, which is associated with the name of Pareto.

Here I would like to look at the hidden correlations, and especially contradictions, within developing human societies undergoing historical change. Though philosophers of history (a Rickert, a Windelband, or a Collingwood) may dislike it, the natural scientist cannot fail to want to try to elucidate the causative elements in history. Hence the profound appeal of the Marxist contributions. The post-Hegelian idea of contradictions lying actually within history, and being resolved as history progresses, seems to be one which has hardly yet reached the beginning of its fruitfulness.

We may glance first at some formulations of contradictions below the social level. The most obvious of these is, at any given stage, the opposition between the old decaying factors and the newly arising factors. As Lucretius wrote:

> ... *omnia migrant*
> *omnia commutat natura et vertere cogit*
> *namque aliud putrescit et aevo debile languet*
> *porro aliud clarescit et e contemptibus exit.*[1]

[1] *De Rerum Natura*, v, 810–33: 'All things move, all are changed by nature and compelled to alter. For one thing crumbles and grows faint and weak with age, another grows up and comes forth from contempt.' Translation by W. H. D. Rouse, Loeb Classical Library ed., p. 399.

And Stalin (or one of his associates) has written that 'internal contradictions are inherent in all natural phenomena; the struggle
between the old and the new, between that which is dying and
that which is being born, constitutes the internal content of the
developmental process.' As the pre-Socratics pointed out, we have
in general, at all natural levels, the forces of repulsion or disaggregation and those of attraction or aggregation. The latter are
usually the newer. They overcome the former but their victories,
though decisive, are never complete; the old repulsive forces
remain incorporated into the new structure. In the architecture of
a protein molecule, for example, there are forces holding atoms
apart as well as the more obvious bonds and valencies which bind
the structure together. So also in the architecture of an embryo,
there are powerful repulsive forces of one tissue against another
('tissue incompatibilities') which have their part to play as well as
the cohesive ones ('tissue affinities') in forming the whole; in recent
years Holtfreter has brilliantly begun the analysis of them. Any
new synthesis at any higher level embodies elements of both the
warring sides at the lower level. This is the secret of all high
levels of organization. Thus mythologically in Greece the Furies
are conducted to their cave under the Acropolis, and in China the
Dragons enrolled in the Civil Service.

Natural processes, as Bernal has pointed out in his *Aspects of
Dialectical Materialism*, are never 100 per cent efficient. Beside the
main process of reaction, there are always residual processes or
side reactions, which, if cyclic, or adjuvant to the main reaction,
will not matter very much. But they may be antagonistic and
cumulative, so that after a lapse of time a new situation will arise
in which an antithesis will oppose the thesis. Then in the ensuing
unstable state of affairs one of the possible solutions may be a
higher level of organization. Bernal applied these ideas to the
formation of planets, the appearance of hydrosphere and atmo-
sphere, and so on. Another discussion from a similar point of view
was that of Haldane in this journal[1] on evolution theory. With
heredity as the thesis and spontaneous mutation as the antithesis,

[1] *Science & Society*, 1937-8, **2**, 473.

variation might be said to be the synthesis. With variation as the thesis, and selection, natural or sexual, as the antithesis, evolution itself might be said to be the synthesis. So also a contradiction could be traced between sexual or other intraspecific competitive selection of the fittest individuals on the one hand, and consequent loss of fitness in the species on the other hand, the synthesis being the survival of those species showing little intraspecific competition.

The early conviction of Engels and Marx (as expressed, for instance, in *Anti-Dühring*, *Dialectics of Nature*, and *Socialism, Utopian and Scientific*) that the whole natural process runs and has run in an objectively dialectical way, was quite rightly directed against the static conceptions of the Victorian scientists, who were unprepared for the mass of paradoxes and contradictions that science was about to have to deal with, as in relativity and quantum theory, and who did not appreciate the overcoming of deadlock antagonisms by the appearance of higher organizational levels. Most of the scientists of T. H. Huxley's time were still dominated by the system of formal logic, challenged though it had been in the very medieval age itself by figures such as Nicholas of Cusa. The well-known rules of the passing of quantity into quality (as in the increasing length of aliphatic hydrocarbon chains or the piling up of static charges until a breakdown of insulation occurs), the unity of opposites (as in the hydrogen ion concentration formulations), and the negation of negations (as in metamorphosis phenomena), have all become commonplaces of scientific thought. This was pointed out by Haldane in his *Marxism and the Sciences*. What the Marxists have not as yet sufficiently done, however, is to elucidate the way in which each of the new great levels of organization (dialectical levels) has arisen. Oparin's *Origin of Life* makes a beginning along this line which has proved widely acceptable in biochemical circles.

One example of a recent scientific advance which bears distinct traces of dialectical thinking is to be found in the new knowledge of muscle contraction. When I was a student just after World War I, our knowledge of the intimate nature of muscle contrac-.

tion was hardly more advanced than it had been at the time of
Borelli's *De Motu Animalium* (1684). But in 1930 came the discovery of von Muralt & Edsall that the particles of the chief
protein of muscle (myosin, a globulin) were even in isolation rod-
or fibril-shaped; and later on the application of X-ray analysis to
muscle by Astbury, Meyer and Mark indicated that the contraction of a myosin fibre, like that of a fibre of keratin (wool) or
cellulose (cotton) was a truly molecular contraction. Only the
contraction of myosin was readily reversible ('ert springiness')
while that of these other fibres was not so ('inert springiness').
Parallel with these discoveries, a large amount of information had
accumulated, as the result of the work of Meyerhof, Embden,
Parnas, D. M. Needham, and others, about the cycles of phos-
phorylation, in which energy from the breakdown of carbo-
hydrate fuel is transferred from molecule to molecule, always
accompanying the transfer of phosphorus, until the substance
adenosinetriphosphate is reached from which the energy passes
directly to the muscle fibril itself. Still, however, there was no
clue as to the relation between the chemical, phosphorylating,
energy-transferring chain of reactions, and the physical, contract-
ing function of the fibrils. In 1937, however, Engelhardt &
Liubimova in Moscow, looking for the enzyme in muscle which
breaks down adenosinetriphosphate and so liberates its energy,
found that this enzyme was none other than myosin, the contractile protein, itself. This discovery having been confirmed by
workers in other countries, the effect of adenosinetriphosphate on
myosin *in vitro* was examined, and it was found that this substance
brings about an immediate shortening of the rod-like myosin
particles, followed by a slow lengthening as the phosphorus is
liberated from the adenosinetriphosphate by enzymic action.
This was reported by J. Needham, D. M. Needham, S. C. Shen,
A. Kleinzeller, M. Miall, & M. Dainty. We thus approach the
conception of a 'contractile enzyme'. The substrate once present,
the enzyme protein has no choice but to act upon it, yet in doing
so its own physical configuration is fundamentally changed. With
the disappearance of the substrate, the configuration change re-

verses. Probably this conception can be extended to explain many things in embryonic differentiation, but here the configuration change would have to be irreversible, just as the contraction of textile fibres is irreversible or nearly so. This set of ideas is reminiscent of the dialectical principle embodied in the line: 'Man is changed by his living, but not fast enough'; and links up with the effect of material techniques upon human thought and human society, to which we must now turn.

So far nothing has been said about contradictions in social development. But in history there is such a wealth of them that it is hard to know what examples to choose. Non-Marxist historians fully recognize them, as for instance did H. Butterfield in *The Whig Interpretation of History*, where he opposed the moralizing attitude to historical conflicts. showing that each side stood for some elements which weɪ ɪmbodied in the subsequent synthesis. The English Civil War in the seventeenth century presents a striking example of a contradiction-deadlock and its solution by the appearance of a new 'phase' of society. The rising merchant middle class, allied with the lesser country gentry, came into sharp opposition with the feudal system represented by the royal court, the aristocracy and the Anglican episcopate. Feudal royalism found its antithesis in the radical puritan republicanism of the Commonwealth, especially in the very advanced ideas of the Levellers (led by Lieutenant Colonel John Lilburne), and the Diggers, or co-operative farmers (led by Gerrard Winstanley), who wanted what today we should describe as a socialist state. Their programme emerges clearly from Winstanley's *Works*, as edited by Sabine, and from the three-volume *Tracts on Liberty in the Puritan Revolution*. But the time was not ripe for their ideas. Cromwell's victory over both the Right (Presbyterians) and the Left (Levellers and Diggers) assured to the merchants of the City of London the controlling power in the Commonwealth as long as it lasted. And the Restoration itself was a synthesis in which a constitutional monarchy (or one which was bound to end explicitly as such) combined with a triumphant parliament controlled by the middle class whose interests the Civil War had made secure. The Protestant *coup d'état* of 1688 and the Reform Act of 1832 simply completed the process.

Other revolutions have shown a similar development. In the French Revolution the feudal monarchy was opposed by the revolutionary Jacobins, and after their victory Baboeuf's group represented the extreme left. But they were quite unable to take the leadership of the movement, and the eventual outcome was the rule of the post-Napoleonic bourgeoisie. So also in the American Revolution, the forces of the left found their expression in men such as Daniel Shay, but the resulting synthesis was the rule of the gentry, leading to capitalism, which, under the exceptional conditions of abundant natural resources in the North American continent, has ruled America, not without success, until today. The dialectical process is perhaps the explanation of the feature so characteristic of revolutions, that they always apparently move 'two steps forward and one step back.' Even the Russian Revolution may not be entirely an exception to this rule. It is true that although, according to previous patterns, the bourgeois Kerensky government should have been the outcome, Lenin's party was able, on account of the further advanced stage of world history, to make the famous 'revolutionary jump' and skip many years of mercantile democracy. But in its early days, many schemes, such as 'moneyless accounting,' attempted complete elimination of nationalist feeling, and the absence of many usual societal restraints, were tried. These have since been abandoned as the growing Soviet society has 'found itself' and attained the colossal strength required not only to survive, but to win, the present World War.

Other examples of hidden correlations and contradictions at the social level are readily found in historical and sociological books, especially those written from a more or less materialist standpoint, and they need not be recapitulated here. Thus the very origin and parallel growth of nationalism with capitalism out of the international, or rather pre-national, Latin-speaking Middle Ages, is itself a contradiction of the strangest kind. The new competitive atomic society of mercantile monads grew up together with the differentiation of the modern European languages—Petrarch and Chaucer no less than Fugger and Gresham moulded capitalist Europe. And ultimately, when we come to modern times, when, aided by the many devices of modern publicity, economic competition rouses nationalism to an insane fervour ending in warfare, the modern nationalist state must arm its workers in the struggle against 'foreign' nationalism or imperialism. Yet this is to arm its own destroyers, as was clearly seen in the Russian Revolution. Rifles and machine-guns are scared to Janus, and the tsarist uniform may be worn with a red armband. In such necessities, such unavoidable consequences, do we not really begin to see some of the causative mechanisms of history at work?

Colonial development, too, embodies contradictions. The typical course of imperialism is presumably to maintain the colonial land as a *région d'exploitation;* that is, to regard its populations purely as consumers for finished goods produced in the homeland, and to remove from it all mineral wealth for elaboration in the homeland or elsewhere. But this state of affairs cannot last. It will not be long before some group of capitalists in the homeland will appreciate the fact that much cheaper labour is available in the colonial country, and they will proceed to erect factories producing consumer's goods within that country, thereby ruining the homeland industry as the Lancashire textile industry was ruined. They will also be joined by 'native' capitalists, probably sprung from the comprador class without which the original conquest of the country could not have been effected. Then, as standards of life and education rise, all the resources of state power will be required to maintain the country in the colonial stage. This will necessitate the raising and arming of 'native' troops. And so the vicious (or rather, beneficial) circle described in the preceding paragraph will start to operate again.

Nationalism, of which we have already spoken, is perhaps a manifestation of individualism; hence its historical association with capitalist monadology. But where individualism is concerned, there is another contradiction. In the beginnings of modern science, individual scientists could make fundamental discoveries, but as science has developed, an ever higher degree of collaboration and co-operation has become necessary, culminating in our own time with the necessity for absorbing individual effort in the efforts of teams. Yet this is difficult because it runs counter to the ethos of capitalist competitive civilization. Hence a paradox which only a change in form of civilization can remove.

One of the greatest contradictions in the capitalist stage of society is no doubt that of education. In the English Civil War, for example, there was no true proletariat, the masses of men employed on either side were comparatively small, and few had enough education or awareness to understand the call of the Levellers and Diggers. The capitalist entrepreneur, theoretically speaking, would prefer to have a reservoir of quite unskilled and unthinking labour power, such as the women and children of the English coal mines a hundred years ago or those of the Yunnan tin mines today. But it is of the essence of developing capitalism that science and technique develop with it. The more far-seeing capitalists support to the utmost that 'improvement of trade and husbandry' which was the slogan of the early Royal Society. Hence there comes more and more the necessity of an educated proletariat; the purely unskilled will not do; besides 'hands', brains and mechanical ingenuity are required. Sir Thomas Gresham, the seventeenth-century English financier, whom we have already mentioned, established by his beneficence three foundations, the Royal Exchange in London, the significance of which as a meeting-place for merchants and early industrialists is obvious; Gresham College in London, where professors of astronomy, navigation, mathematics, and all 'useful arts' taught the promising apprentices and young ingenious merchant-venturers; and finally Gresham's School at Holt in Norfolk, for the same training for younger children. But so, as time goes on, there grows up a vast class of technicians parallel with the unskilled labourers, until in our own day, in spite of belt assembly lines, machine-minding, and other devices, unskilled labour threatens to disappear altogether, leaving the technicians in the field. These, however, can never accept the status of pure wage-slaves to which theoretical capitalism would like to condemn them. Hence the inevitable transition from capitalism to some such economy as that of the Soviet Union, where the dignity of each man's work is fully recognized, and his participation in the government of the factory and the state is acknowledged.

Naturally capitalist society objects to this transformation; hereditary power and privilege cannot be expected to give up their possessions without struggle. So in the last resort Nazi-fascist theory is brought in to save the decaying structure. For Reason, always the banner of rising social classes, it substitutes a fantastic irrational mythology, in terms of such ideas as 'the mystical blood-substance of the Germanic race'; the 'mystical embodiment of the wills of individuals within the heaven-inspired will of the Leader', the 'divine mission of the Germanic or Japanese people to dominate all other peoples', and so on. But on the other hand modern capitalism, even when profits are guaranteed and strikes abolished under Nazi-fascism, cannot get on without effective control over Nature, whether in peaceful or in war-making mechanisms, and this absolutely necessitates scientific rationality. Hence a schizophrenia in the Nazi-fascist-Shintoist social structure. The Fascist metal-worker, measuring accurately to a thousandth of a millimetre, must believe that his Slavonic counterpart is a fiend in human shape. The Shinto-fascist bio-chemist, preparing emergency rations from silkworm protein, must have faith that his emperor is really descended from the sun. The Nazi radio technician must believe in the literal inspiration of Hitler's every action, a belief more and more difficult to entertain the more defeats he suffers. Ultimately only a rational society can command the allegiance of scientific technicians. This is the point at which we come to the relations of science and democracy.

II

Perhaps it is a commonplace to say that science is only possible in a
democratic medium. But commonplaces may be quite erroneous
when carefully examined, and if this one be true, it cannot be
thought of as established in the absence of detailed consideration.
I believe, however, that there is a fundamental correlation between
science and democracy, one of those hidden correlations which are
the obverse of the contradictions already discussed.

In the first place, it is quite clear, and the fact is not contested by
any scholars, Marxist or otherwise, that historically science and
democracy grew up in western civilization together. I refer here,
of course, not to primitive science, nor to medieval empirical
technology, but to modern science, with its characteristic combination of the rational and the empirical and its systematization of
hypotheses about the external universe which stand the test of
controlled experiment. To the English Civil War and Commonwealth period we have already referred, but this was only part of
that great movement lasting from the fifteenth, perhaps, till the
eighteenth century, in which feudalism was destroyed and
capitalism took its place. Other aspects of the same change were
the Protestant Reformation, the literary Renaissance, and the rise
of modern science.

Exactly why modern science should have been associated with
these changes still remains to be fully elucidated. Here it may
suffice to say that the early merchant-venturers, extremely far-
seeing men, and the princes who supported them and based their
power on them, were interested in the properties of things because
on such properties alone could mercantile and quantitative economics develop. At no time in history has scientific research been
possible without financial support. In antiquity the aid of princes
had to be obtained; Babylonian astronomy was a department of
state; Greek physics and mathematics relied on the support of the
sovereign city-states; and Alexandrian biology depended on the
Ptolemies. In the first beginnings of capitalism the merchants and

embryonic industrialists provided the essential funds for the experiments of a Boyle or a Lavoisier, however indirect might be the channels. In the Restoration Court, after the victories of the Commonwealth, scientific and technical experiments were performed under the very auspices of King Charles II; 'the noise of mechanick instruments resounded in Whitehall itself'. Business had, in fact, become respectable, and science with it. The younger sons of the aristocracy hastened to apprentice themselves to thriving mercantile, mining, and industrial enterprises. Waning shadows of the former feudal exclusiveness appear in the attractive story of John Graunt, the founder of vital statistics and the first man to apply mathematical methods to the 'Bills of Mortality'. The Royal Society were uncertain whether to admit him of their number, since he was some kind of small tradesman in the City, and they sought the opinion of the King on the matter, who replied that 'they should certainly admit Mr Graunt, and if they found any more such tradesmen, they should be sure to admit them also without delay'.

This judgment was what we could call the affirmation of a democratic principle, namely that in matters of science and learning birth or descent is a thing indifferent. And indeed it is obvious that if the rising merchant class encouraged the sciences to develop, their slogan on the political side was precisely democracy in all its forms. Everyone knows, of course, that democracy, *sensu stricto*, took its origin in ancient Greece, but it was always based there on a helot or slave population, and it seems extremely unlikely that it would ever have developed as it did if it had not been for Christian theology, with its emphasis on the importance of the individual soul before God. Those who took the individual soul seriously were bound to end, as authoritarian thinkers had always realized, by taking seriously the opinions of its owner, not only on spiritual matters, but ultimately on temporal ones too. In 1641 Edmund Waller, one of the Right in Parliament, hit off the situation perfectly:

'I look upon episcopacy as a counterscarp or outwork; which, if it be taken by this assault of the people, and withal, this mystery once revealed, that we must deny them nothing when they ask in troops, we may, in the next place, have as hard a task to defend our property as we have lately had to recover it from the royal prerogative. If, by multiplying hands and petitions, they prevail for an equality in things ecclesiastical, the next demand perhaps may be for the like equality in things temporal. I am confident that, whenever an equal division of lands and goods shall be desired, there will be as many places in Scripture found out which seem to favour that, as there are now alleged against the prelacy or preferment of the Church. And as for abuses, when you are now told what this and that poor man hath suffered by the bishops, you may be presented with a thousand instances of poor men that have received hard measure from their landlords.'

A most revealing passage. Waller represented the middle class which did, in fact, win the Civil War, and founded the eighteenth and nineteenth centuries, recovering its property (essential word) out of the power of fuedal kingship, but not surrendering it to the socialist state which the Levellers and Diggers would have wished to see.

Yet democracy was the natural slogan of the 'bourgeois' revolutions. If the Left interpreted it in a manner too thorough-going, it mattered little, as the real strength lay with the centre. Nevertheless, especially in the earlier phases of the Civil War, this slogan was an essential rallying cry for the forces which helped the middle class to conquer, and were thereupon themselves temporarily liquidated. About the year 1648 a preacher of the Right was inveighing one evening in the pulpit at Christchurch in Oxford against the indiscriminate preaching of the New Model Army's troopers. When he came down from the pulpit, there stood a polite young man, clad in the red coat of an officer of foot, who spoke to him somewhat as follows: 'You speak against the preaching of the soldiers, but bethink you that if they cannot preach, then will they not fight, and then what will become of all of us, and of that Army which hath been these last few months so helped by God that it hath captured twenty-five great cities, won five great battles, and I know not what besides.'

There can be no doubt that capitalism, democracy, and modern science grew up together. The question arises whether the two latter are essentially dependent upon the former. Many observers today consider that capitalism has ceased to be the matrix for the other two which once it was. There is no space here to estimate the arguments involved, but from all indications it certainly does not seem as if science has suffered in the only state, Soviet Russia, which has abandoned the capitalist system and proceeded to socialism, rather the contrary. A greater proportion of the national income is devoted to science there than in any other country, as is indicated by the figures in Bernal's *Social Function of Science*. And as far as democracy is concerned, it is likely that there are other means of expressing it besides those traditionally associated with parliamentary representative systems. Such is the tenour of the Webbs' study, *Soviet Communism*.

But let us turn now to the philosophical, rather than the historical, connections between science and democracy.

In the first place, Nature is no respecter of persons. If someone takes the floor before an audience of scientific men and women, wishing to speak of observations made, experiments carried out, hypotheses formed, or calculations finished, the status of this person as to age, sex, colour, creed, or race, is absolutely irrelevant. Only his or her professional competence as observer, experimentalist, or computer is relevant. The community of co-operating observers and experimenters upon whose activity science is based is fundamentally democratic. It surely prefigures that world co-operative commonwealth which we see as the inevitable crown of social evolution. The admission of Mr. Graunt to the Fellowship of the Royal Society symbolizes the absolute equality shared by all competent observers of nature. Just as the theologians of patristic Christianity emphasized the importance of each individual soul, so the leaders of modern science at all its stages have recognized the importance of each individual observer. They may, in practice, have given greater credence in any particular age to those whose merit as scientific workers was widely recognized, but the history of science now contains so many examples of the neglect, or persecution, of individuals who afterwards proved to be profoundly right (such as Galileo in astronomy, Mendel in genetics, Semmelweis in pathology, Willard Gibbs and Wollaston in physical chemistry, and many others), that today no one is likely to fail to obtain a fair hearing. A competent observer is a competent observer, no matter whether his genes have given him the pink and white skin-colour of the Euro-American, the golden-brown of the Chinese, the blue-black of the Negro, the dark brown of the Indian, or even the white of the Albino. Greeks and Arabians, Jews and proselytes—everyone remembers the famous passage: 'Men of Cappadocia and Pamphylia, and from the parts of Libya about Cyrene; we do hear them speaking in our tongues the mighty works of God.'

If science is democratic as regards 'racial' characteristics, it is also democratic as regards age. In science it is so very easy to be wrong that no scientific worker, no matter how great his age or experience, can afford to despise the contribution of some younger man, even if comparatively inexperienced. It has recently been well argued by Waddington in *The Scientific Attitude* that science solves the old problem of authority and freedom in a remarkable way. Authority is safeguarded because the structure of the scientific world-view is based securely on the results of the co-operative labours of a million investigators and is not therefore easily over-thrown. But at the same time freedom is provided for in that it is always open for any individual thinker to insist upon its modifica-tion, if he can. In the case of an Einstein, the modifications may be fundamental. Were human society organized in a similarly rational way, were there no governing class with vested interests requiring irrational preservation, a similar solution of the problem at the social level might be found.

One corollary of the indifference of science to age is that those civilizations which have developed an exaggerated veneration of the old, and an exaggerated respect for teachers, will have to modify it in taking their place in the modern world, just as western civilization overthrew the authority of scholastic Aristo-telianism.

We may also ask what is the estimate of man to which three centuries of science have led. It is certainly not that estimate which the totalitarian philosophers of our time have formed. Men are not naturally machine-minding slaves, fit only to carry out the undisputed orders of self-appointed leaders in some kind of hierarchy; they are basically rational beings, each with his right to life, love, labour, and happiness, each with his distinctive contribution to make to the well-being of the community, each with a right to an opinion and its expression on the community's form, laws, and actions, each with a duty to guard the community's property and to uphold its just laws. As Jennings has so well put it in *The Biological Basis of Human Nature*, the type of human community to which the facts of biology most clearly point is that of 'a democracy which can produce experts'. Nazi-Shintoist-fascist racialism which purports to have a biological foundation is the greatest scientific fraud yet perpetrated on a public insufficiently critical to detect it.

But what are the experts to whom Jennings refers in the above passage? They are certainly not armchair, office, or library penholders and bookworms, burrowing among the opinions of antiquity to find something useful for the needs of today. They are men with practical experience in the laboratory and at the engineer's bench, in the dyeing-sheds, the steel mills and the hospital wards. It is here that we come upon yet another hidden connection between science and democracy, namely the bridging of the gulf between the scholar and the artisan.

This antithesis has existed throughout history; one can find traces of it in the earliest civilizations. From the very beginning the philosophers and mathematicians and primitive scientists allied themselves, together with all those who were able to master the arts of reading and writing, with the scribes and civil servants who carried out the orders of the government of kings and nobles. The artisans, such as those who understood chemical and smelting processes in ancient Egypt, or the murex dyers in later Roman times, or the jade-cutters of China or the Aztec makers of feathered robes—all were set apart as belonging to the inferior 'mechanic' class with whom gentlemen would not associate. Hence the complete gulf between theoretical scientific thought and technical human practice for thousands of years. Aristotelian and Epi-

curean speculations such as the atomic theory, went their way
without any influence on actual techniques; while the Chinese
scholars, masters of the ideographic characters, but remote from
their own technical artisans, continued until a late date to harp on
the primitive theories of the five elements and the two principles;
the *yang* and the *yin*. Indeed the Greek distinction between theory
and practice, the former suitable for a gentleman and the latter not,
finds its exact counterpart in the Chinese distinction between
hsüeh and *shu*. Only some of the greatest figures broke through
these boundaries, as for example Aristotle when he conversed with
fishermen and shepherds before writing his immortal *De Genera-
tione* and *Historia Animalium*, or Hippocrates the practical
physician, or Archimedes the practical military engineer; or in
China, Shen Kua who in the Sung dynasty abandoned Confucian
exclusiveness to record with care all natural wonders, such as for
the first time the magnetic compass, or the Taoist outcast scholars,
whose alchemy penetrating to the west was the foundation of all
modern chemistry.

With the coming of capitalism, however, all was changed.
Feudal disinclination to soil one's hands by practical manual
operations went out of fashion for ever. Now the merchants set
the tone. It was now not only a matter of handling exploitable
commodities in their warehouses, of examining the properties of
oils and waxes, animal and plant products, minerals and metals.
The time had come to investigate all such investigatable things to
the bottom, and as part of this enterprise, a new interest awakened
in the traditional arts and trades which since time immemorial
had been gradually developing through the labours of a million
empirically-minded artisans. The beautifully illustrated books of
the eighteenth century describing all the trades and husbandries
came into being as a result, in both the Orient and Occident.

This new impulse was profoundly democratic. Henceforward
continuously the scholar and the artisan were to meet and ultimately to mingle, until in our own time they find a perfect fusion
in the great experimentalists, a Pavlov or a Langmuir, working
indifferently with brain and hand. The scholar, from being the

support of the king, becomes the comrade of the artisan. Hence perhaps, that slight tension which one sometimes notices between scientists and lawyers. Lawyers frequently represent the scholars of ancient times, those who spend their time propping up the existing order. Eggleston, himself a lawyer, has written in his *Search for a Social Philosophy*: 'Whatever may be said to the contrary, it seems to me that the primary function of law is to put a brake upon the occurrence of inevitable change. Though this may no doubt be necessary, I am therefore unable to agree with those who would have us regard the law as a sublime principle.' Scientists, on the other hand, are prominent among the factors of social change, though they cannot always as yet control it. They are certainly not among the factors propping up existing orders.

On the psychological side also there are powerful hidden connections between science and democracy. There is a real kinship between the scientific mind and the democratic mentality. There is in both cases a basic scepticism—the scepticism of the experimentalist is mirrored in the scepticism of the voter. The Royal Society's motto *Nullius in Verba*: take nobody's word for it; see for yourself; is also that of government by and of the people. Though it may be a counsel of perfection, it sets the psychological key. Making one's own mind up on the available evidence, deciding for oneself what goal to aim at, weighing the facts from a dozen different quarters, these are characteristics both of the scientist investigating nature, and the democratic citizen taking part in the governance of the State. They do not let others decide on their aims, they do not leave to others the assessment of evidence, blindly believing in the superior leadership of particular men. There is a quality of openness to conviction, of give and take, a live and let live attitude, characterizing both the scientific worker among other scientific workers, and the democratic citizen among his fellows. Applying the rules, too, when no one else is about, does this not also unite the scientific worker and the citizen? Even in very recondite studies, where it is unlikely that he can be checked in his own lifetime, the scientific worker is on his honour to make no falsification of results; he cannot know what might depend on it. So also (as I once thought when observing a driver threading his way tortuously through traffic round-abouts in the middle of the night, with no other vehicle moving for miles around), the democratic citizen adheres to the rules in the absence of all usually present restraints. He does so because in his inmost self he admits them to be good and rational. The laws of a rational society are written in the hearts of its citizens as well as in its codices.

Then objectivity. Authoritarian theory carries over into all life the mechanical system of military discipline. But Nature is not amenable to that. Objective facts are objective facts, irrespective of the leaders' wills. As we have seen in Nazi Germany, scientific anthropology, for instance, may be broken, it may be exiled, but it will not bend. So also in democratic thought. Subjective valuations may flourish for a time, but before long the hard facts of personal well-being or personal misery will indicate to the citizen with indubitable clarity where his interests lie. The subjective and the irrational are anti-democratic, they are the in.. 'ments of tyranny.

Authoritarian, tyrannical thought, again, tends always to divide and to keep separate, yet the fundamental urge of science is to form a unified world-picture, in which all phenomena have their proper place. In Thomson's *Aeschylus and Athens*, a work dealing with the anthropological origins of Greek culture and folklore, and the rise of Greek drama and literature from them, this cor-relation is clearly made. An examination of Ionian science (especially Anaximander's contribution) and Orphic mystical theology, suggests that 'the tendency of aristocratic thought is to divide, to keep things apart,' while the 'tendency of popular thought is to unite.' In Orphism, love implied the reunion of what had been sundered. Divisiveness must have followed naturally from the efforts of the earliest aristocrats to maintain separate from the communal land the special enclosure or *temenos* which they had won by cunning or bravery. Maintenance of differences has been throughout history, even when the differences were quite illusory, the basic aim of all tyrannical, aristocratic, or oligarchic social orders. Conversely union and withness, the *ta thung* of Chinese thought, has been the aim of democracies. And since, as we have seen above, the principle of union, attraction, and aggregation persistently triumphs, not only in the living world, but in the non-living world also, throughout the evolutionary process, over the principle of disunion, repulsion, and disaggregation, it is clear that the democratic trend at the human social level has the future before it while the authoritarian trend is doomed.

Democracy might therefore almost in a sense be termed that practice of which science is the theory.

III

Thus we may summarize the hidden connections between science and democracy. By no coincidence, modern science and democracy arose together during those stormy centuries during which modern society was born. Science is profoundly democratic in its relation to differences of race, sex, and age. It finally solves the ancient problem, so vexatious for theologians, of authority and freedom. Its estimate of human social order is that of a democracy which can produce experts. And these experts finally bridge the age-old gulf between the scholar and the artisan, uniting in single persons the highest human attainments of hand and brain. Lastly, the psychological attitude in science and democracy is very much the same. Both are on the side of union, attraction, and aggregation, leading to the higher organizational levels.

If, then, there is a connection between science and democracy, the opposition must come between science and all possible forms of irrational arbitrary tyranny, authoritarianism and totalitarianism. And the contradiction today is that capitalism, in its late highly developed monopoly form, needs science more than ever, for without it the hitherto unapproached technically high standard of life cannot be maintained or extended. Yet since capitalism is ceasing to be able to appeal to mankind on rational grounds, it more and more tends to have recourse to the irrationalities of authoritarianism. Hence, though capitalism and democracy grew up together, we approach the time when either one or the other of them must go. And since science is indispensable to all future human civilization, it is capitalism that is doomed, and not democracy.

Supposing, therefore, that these arguments are correct, might we not hope to see some justification of them during World War II, through which we have been painfully living? The Axis powers began the war with an enormous handicap in their favour in the application of the sciences to warfare. The principle of *Wehrwissenschaft*, of science valued only for its applications to warfare, had been adopted by the Nazis, the fascists, and the Shinto-fascists alike. In the early days of the war the world was amazed by the efficiency of the Nazi war machine, in which every possible contingency seemed to have been thought out beforehand. Yet as the years passed by, how did it go on every front? It is not presumptuous, but merely objective, to say that on every front, and in every theatre of war, on land, on sea, and in the sky, the technology of the democracies has proved superior to that of the fascist powers. Even in the Battle of Britain, it was the design of the British planes, no less than the valour of the British pilots, which decided the matter against the Nazi air force. Later on, in the amphibious operations so carefully studied by the democracies, and used with such success by the Americans in the Pacific no less than by the allied expeditionary forces in North Africa and France, democratic technology outdid that of the Axis. The Nazi-fascist forces were not turned out of Africa itself without high technical endeavour, for they themselves had prepared with the utmost care for desert conditions. And so, all along the line, with a host of special inventions, such as radio-location devices, fog-piercing lights, radio beacons, television locators, bridging machinery, inter-tank communications, tank-carrying gliders, rocket propulsion, explosion theory, gas turbines, turbo-superchargers, gunnery predictors and calculating machines—in few cases have the democracies not shown themselves superior to all the fascist powers. Even new branches of science have been developed, such as terminal ballistics and mycological antiseptics; while others, such as aeronautics and applied protein chemistry, have made steps forward. In all of this application of science to war on behalf of the democracies, the part played by Soviet scientists and technicians was certainly a large one; its importance will be recorded when the technological history of the war is written.

Nor must we forget the negative aspect of all this. New weapons and devices are important, but there is also the countering of those which the enemy introduces; his new means of offence must be offset by new defences. Sometimes, as in the case of the magnetic mine, such new defence measures have to be found very quickly. Here again the scientists of the democracies were not wanting, and when the time comes it will be possible to unfold many a heroic tale concerning those, for example, who investigated the secrets of unexploded bombs, time-action land-mines, and similar devices. Lastly, besides new weapons and the means of defence against new weapons, there is the most efficient utilization of those already possessed, and here again the younger scientists of the democracies, acting in the field as 'civilian scientific officers' carried out and interpreted the necessary experiments. All in all, this war will have clearly demonstrated the superiority of science in the democracies.

We might put it this way. The Axis powers have carried out a great social experiment. They have tested whether science can successfully be put at the service of authoritarian tyranny. The test has shown that it can not.

IV

Let us now turn to consider a special case, a highly complex set of phenomena at the social-historical level, in which, as it happened, I became interested on account of war service in China. It is the problem of the origin of modern science in Semitic-Occidental (Euro-American) civilization on the one hand, and its failure to arise in Chinese civilization on the other. From what has gone before, it might be possible to say baldly that if science and democracy have been so closely connected, we need look no further for the failure of science to arise in China. But this, though it contains some truth, is a distortion of the facts, since Chinese civilization contained many democratic elements, though they differed in form from those of the west. It is the intention of the present writer, after the war, to set forth the following argument in fully documented form; here only an adumbration of it can be given. That the problem is worth investigating can hardly be doubted; indeed, it seems one of the greatest problems in the history of civilization, since no other culture, except the Indian, equals that of the West in scope and intricacy more closely than that of China.

One may begin by denying the accuracy of the title adopted by the philosopher Fêng Yu-Lan in a famous essay: 'Why China has no Science.' China had an abundance of ancient philosophy, and of medieval science and technology; only modern science (until the coastwise impact of the West) was lacking. The first thing to do is to establish this fact, and it is not difficult. In ancient philosophy I would say that the Chinese were as well able to speculate about Nature as the ancient Greeks, on whom is usually laid all the kudos for the establishment of the scientific world-view. Not, of course, the Confucian school, which was only interested in social justice, but among the Taoists (Lao Tzu, Chuang Tzu, Kuan Tzu), the Mohists (*Mo Ching*) and the Ming Chia or school of logic (Kungsun Lung), there are thoughts about Nature quite as profound as anything in Greek speculation. In practical matters, too, the Taoists (it is now established) were the originators of alchemy, which was probably brought to Europe through the Arabs; while the Mohists exhibited a strong interest in physics and optics. All this activity was going on during the first five centuries B.C. parallel with Ionian and Attic science in Greece. Nor did scientific philosophy cease with the beginning of the Christian era. In the late Han dynasty we have Wang Chhung, one of the greatest sceptics of any country, and in the Sung dynasty the whole school of so-called Neo-Confucians headed by Chu Hsi, who has been sometimes termed a pre-Victorian Herbert

Spencer, and whose philosophy is through and through natural-istic. He it was who first appreciated the true nature of fossils, 400 years before Leonardo da Vinci in the West. And at the turn of the Ming and Chhing dynasties there was Wang Chhuan-Shan, who has been described as a Marxist before Engels.

All this demonstrates that the Chinese were perfectly able to speculate about Nature. That they were perfectly able to carry out experiments along empirical lines further appears from the triumphs of their medieval technology. It is widely appreciated that their inventions changed the course of world history. Without paper, printing, the compass, and gunpowder, how would the change from feudalism to capitalism in the West have been pos-sible? Paper in China goes back to the first century A.D., and printing to the eighth, quickly attaining to movable block printing in the eleventh. The magnetization of the needle by the lodestone was known long before then, but the first statement of its directive property (as well as its declination) was made by Shen Kua in the Sung dynasty at that time. Gunpowder, probably a product of Taoist alchemy, first appears in the late Thang time, but is used for warfare in the Wu Tai period (tenth century A.D.) four centuries before its similar use in the West.

Besides these inventions and discoveries, the effects of which had so powerful an influence elsewhere, there were others no less in-teresting. Here one can only mention a few: the knowledge of deficiency diseases and a kind of vaccination in the Yuan dynasty (fourteenth century); the technologies of porcelain and silk, grad-ually improving over more than a thousand years; the origination of special agricultural arts, such as the cultivation of oranges and the preparation of natural plastics such as lacquer and of special products such as insect wax; the development of metallurgical procedures and the use of wrought iron chains for suspension bridges, the invention of pound-locks on canals, and the growth of an indigenous pharmacopoeia, some of the drugs in which have valuably contributed to the world armamentarium of today.

Freely one may grant that after a certain time was passed, Chinese technology progressed very slowly, while Western technology, inspired by the appearance of modern science, far outstripped it. The famous dictum that the Chinese peasant was ploughing with iron ploughs when the European peasant used wood, but that he continued to plough with iron ploughs when the European farmer had begun to use steel, is true. In anatomy, for instance, it is likely that the famous 'Fünfbilderserie' of illustrations came originally from China. In the Thang dynasty it was ahead of anything in Europe, but such illustrations continued to be re-produced, mainly in books on forensic medicine for the use of magistrates, in the Chhing dynasty, long after European anatomy had passed forward out of sight. So also in chemical industry. Anyone visiting the famous brine workings at Tzu-liu-ching in Szechuan might imagine that he had stepped into the pages of Agricola's sixteenth-century *De Re Metallica*, or rather its Chinese equivalent the *Thien Kung Khai Wu*, so faithful to tradition are the methods employed.

So we come to the fundamental question, why did modern science not arise in China? The key probably lies in the four factors: geographical, hydrological, social and economic. All explanations in terms of the dominance of Confucian philosophy, for instance, may be ruled out at the start, for they only invite the further question, why was Chinese civilization such that Confucian philosophy did dominate. Economic historians such as Wu Ta-Khun, Chi Chhao-Ting and Wittfogel, tell us that though Chinese and European feudalism were not unlike, when feudalism decayed in China, it gave place to an economic and social system totally different from anything in Europe: not mercantile, still less industrial, capitalism, but a special form which may be called Asiatic bureaucratism, or bureaucratic feudalism. As we have already seen above, the rise of the merchant class to power, with their slogan of democracy, was the indispensable accompaniment and *sine qua non* of the rise of modern science in the west. But in China the scholar-gentry and their bureaucratic feudal system always effectively prevented the rise to power or seizure of the State by the merchant class, as happened elsewhere.

On the geographical side, China is a continental country,

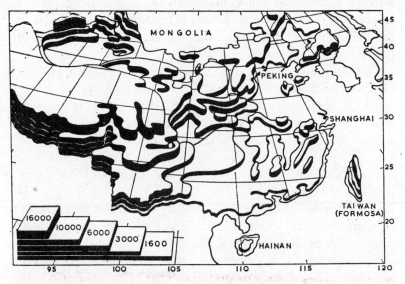

FIGURE 20 Diagrammatic map of China (after Lo Kai-Fu) to show the continental character of the amphitheatre facing the Pacific Ocean to the East and
backed by the Tibetan massif to the West. The issue routes of the Yellow River
and the Yangtze can be seen, with the basin of Szechuan west of the gorges.
Latitudes and longitudes give the scale, with altitudes in feet.

bound to agriculture, not an archipelago teeming with maritime
mercantile city-states, as Europe always was. China's rainfall,
moreover, is highly seasonal, and highly variable from year to
year; this necessitated works of irrigation and water-conservation,
river control, drainage and inland navigation, far greater than
anything which had to be done in the West. Some have seen the
origin of the Asiatic bureaucracy in the necessary control of the
masses of manpower which were required to construct these great
works, but perhaps it is more probable that an irrigation-civiliza-
tion impinged upon feudalism because the great engineering works
transcended the boundaries of the lands of individual feudal lords.
Their power thus inevitably became absorbed by the imperial
power, and after the union of the country under the first Chhin
Emperor in the late third century B.C., the feudal lords steadily
declined until they lost all importance. Correspondingly the

imperial bureaucracy gained in importance, and while on the one hand it prevented any recrudescence of feudalism, it also absolutely prevented the rise of the merchants and the coming into being of capitalism.

This is the background, then, which alone enables us to say that there was no modern science in China because there was no democracy. Democracy of a sort there was, in so far as (in many dynasties at any rate) it was possible for a boy of whatever origin to become a great scholar (the village neighbours might club together to provide a tutor for him) and so take a high place in the official bureaucracy. Democratic, too, was the absence of hereditary positions of lordship, and democratic was, and still is, the psychological attitude of the commons within whom the four 'classes' (scholars, farmers, artisans, and merchants) interchanged with considerable fluidity among one another. It explains the utter lack of the servility so noticeable in other peoples of the eastern hemisphere. But that particular sort of democracy associated with the rise of the merchants to power, that revolutionary democracy associated with the consciousness of technological change, that Christian, individualistic and representative democracy with all its agitating activity, which characterized the New Model Army, the Army of the Marseillaise, the Minute Men, the Floating Republic, the Dorset Martyrs, the Communards, the Sailors of Invergordon and Kronstadt, and the Motor-Cycle Battalions which took the Winter Palace—that China never knew until our own day.

In a nutshell, my view is that in Chinese civilization there were factors inhibitory to the growth of modern science, while in Western civilization, the factors were favourable. It may not be too much to say that had the environmental conditions been reversed as between Euro-America and China, all else would have been reversed too—all the great names in the heroic age of science, Galileo, Malpighi, Vesalius, Harvey, Boyle, would have been Chinese and not Western names; and in order to enter today fully into the heritage of science, Westerners would have to learn the ideographic script just as the Chinese now have to learn alphabetic languages because the bulk of modern scientific literature is written in them.

V

A few concluding words. Science is a function of the socially acting human mind. Science reveals to us the levels of organization in the universe. But both mind and science are themselves products of the social level, and may be studied like any other phenomena at the level appropriate to them. This essay has raised a number of topics, the concept of levels of organization itself, the problem of the dialectical transitions between them, the correlations and contradictions at the social level, the relationships of science and democracy as social level phenomena, and the special case of their appearance in the western rather than the eastern hemisphere.

The aim was simply to try to reveal some of the causative elements in history, the underlying contradictions which make inevitable a certain historical development; to try to answer why science appears at one place and time rather than another, and why it consorts with one social system rather than another. If this task was too presumptuous for a biochemist and embryologist, the duty of a free citizen, and experiences, unexpectedly intimate, of another civilization in another hemisphere, no less loved than his own, may serve as some excuses.

4

SCIENCE AND SOCIETY
IN ANCIENT CHINA

Conway Memorial Lecture, delivered at Conway Hall, London, on 12 May, 1947; revised for *Mainstream*, 1960, **13** (no. 7), 7.

What I want to do in this address is to try to sketch a sort of pattern of the organization of Chinese society, in the course of which a number of points will come up of interest to an audience naturally absorbed in problems of rationalism, ethics, and religion in social life. I am led to do this because, in my thinking on such subjects, I am always working towards a study of what I believe is one of the greatest problems in the history of culture and civilization— namely, the great problem of why modern science and technology developed in Europe and not in Asia. The more you know about Chinese philosophy, the more you realize its profoundly rationalistic character. The more you know of Chinese technology in the medieval period, the more you realize that, not only in the case of certain things very well known, such as the invention of gunpowder, the invention of paper, printing, and the magnetic compass, but in many other cases, inventions and technological discoveries were made in China which changed the course of Western civilization, and indeed that of the whole world. I believe that the more you know about Chinese civilization, the more odd it seems that modern science and technology did not develop there.

To begin with, I would like to say something about the origins of civilization in China; which means the origins of Chinese feudalism, growing up from about 1500 B.C. One must remember that it was always very distinct from the other great civilizations. We know that the river-valley civilizations of Mesopotamia and Egypt were closely linked together from an early date and, similarly, that the ancient civilization of the Indus valley had its connections with Babylonian civilization. The only great river-valley culture which did not have a close connection with these was the Yellow River civilization, that of the Huang-Ho, which became the cradle, especially in its upper regions, of the Chinese people. Actually, as I want to emphasize in a few minutes, that civilization was linked by a number of strands with the Bronze Age in Europe. In spite of this, however, the Yellow River civilization was more independent than connected with the West.

The origins of this first form of Chinese society are very important, because one can see that Chinese philosophy goes right back to them. Great scholars like Granet, the French sinologist, have demonstrated that the origin of towns in China was probably connected with the beginning of the working of bronze, no doubt because the first metallurgists had to have installations of some complexity which required protection from the changes and chances of life in the villages of the primeval tribal community. Granet has traced the way in which the primitive pre-feudal society gave way to the feudal society of the towns of the full Bronze Age in China.

For example, we know that many of the poems contained in the *Shih Ching*, the famous 'Poetry Classic,' are ancient folk-songs. They still show us today something of the songs which were sung by the bands of young men and girls, dancing in those ancient reunion festivals at spring and autumn at which the process of mating was accomplished; the people coming together from their villages to these meetings, these fairs of spring and autumn. The first feudal lords captured the holiness of these places where the people congregated and transferred it to the sacred mound or temple of the feudal 'State' in the town which was then first originated. During what we may call the high feudal period in China, which runs roughly from the eighth century to the third century B.C., the feudal lords were assisted and counselled by a group of men who afterwards became the school of philosophers which we know as the Confucian School.

The Confucian philosophers originated, then, as the counsellors of the feudal lords, and the chief characteristic of that school (not only Khung Fu-Tzu himself but his great disciples Mêng Tzu and later on Hsün Tzu and many others) was a rationalist, ethical approach, embodying a profound concern for social justice as the Confucians understood it. There are many stories about Confucius which I might mention to you. Just by way of example, on one occasion, when Confucius was travelling in a chariot and wanted to cross a river, he and his disciples could not find the ford. He therefore sent one of them to consult with some hermits nearby asking for information as to the way across. The hermits, however, gave a sarcastic answer, saying: 'Your master is so wise and clever, he knows everything, and must certainly know where the ford is.' Confucius was sad when this was reported to him and said: 'They dislike me because I want to reform society, but if we are not to live with our fellow men with whom can we live? We cannot live with animals. If society was as it ought to be, I should not be wanting to change it.'

The general characteristic of Confucian philosophy was thus entirely social—a feudal ethic, no doubt, but extremely social-minded. The Confucians were quite convinced of the need to organize human society in such a way as to afford the maximum of social justice under feudal custom, and they were determined that it should be so organized. They differed, therefore, from other philosophical schools which were not interested in human society nor in how it should be organized. These hermits to whom I have referred may well have been early representatives of the school of thought which afterwards became known as Taoism. I suppose the two greatest currents in Chinese thought are the Confucian on the one hand and the Taoist on the other.

The Taoists were those who professed to follow a 'Tao' and, by this expression, 'the Way,' there is no doubt that they meant the Order of Nature. They were interested in Nature, whereas

FIGURE 21 Illustration from an early Ming edition of the *Tao Tê Ching*, showing Li Tan riding away into the West and meeting the frontier official Kuan Yin.
Though both are quasi-legendary characters, the *Tao Tê Ching* is attributed to
the former, Lao Tzu, and a later book of philosophy, the *Kuan Yin Tzu*, to the
latter. On this incident Berthold Brecht wrote a memorable poem, here in part
quoted:

> 'When he was seventy and growing frail
> The teacher after all felt the need for peace,
> For once again in the country kindness did not prevail
> And malice once again was on the increase.
> So he tied his shoe-lace . . .
>
> But before the fourth day's rocky travelling was done,
> A customs man interposed his authority.
> "Please declare your valuables" — "None".
> And the boy leading the ox said, "A teacher, you see."
> That met the contingency.
>
> But the man, cheerful, and struck by a sudden notion,
> Went on to ask " Who discovered something, you'd say?"
> The boy replied, "That yielding water in motion
> Gets the better in the end of granite and porphyry.
> You see, the hard thing gives way" . . . '

tr. Michael Hamburger.

the Confucians were interested in Man. One might say that the Taoists felt in their bones, as it were, that until humankind knew more about Nature it would never be possible even to organize human society as it should be organized. The Taoists have left us a number of very important and profound texts, among which the famous *Tao Tê Ching*, 'The Canon of the Virtue of the Tao,' is one, and the writings of some philosophers such as Chuang Tzu, who has been considered the equal in his way of Plato. We have these writings still, perhaps in more or less distorted form, like all ancient writings, but in a form in which the thought can still be followed.

The Taoist hermits, who withdrew from human society in order to contemplate Nature, did not, of course, have any scientific method for the investigation of Nature, but they tried to understand it in an intuitive and observational way. If their interest in Nature was such as I am suggesting, we ought to find they were associated with some of the early beginnings of science. And that is in fact the case, because the earliest chemistry and the earliest astronomy in China have Taoist connections. It is now well recognized that alchemy—which we may call the search for the philosopher's stone, and the drug or pill of immortality—goes back well into, and even beyond, the earliest imperial period in China. One of the earliest references to it occurs in the time of the Emperor Han Wu-Ti about 130 B.C., in which the magician Li Shao-Chün goes to the Emperor and says: 'If you will sacrifice to the stove, I will show you how to make vessels of yellow gold and from these you may drink and achieve immortality.' That is perhaps the earliest record of alchemy in the history of the world, and sacrificing to the stove is equivalent to someone saying to-day: '*If* you support my researches, I will, etc.' In the second century A.D. there is on record the earliest book known in the history of science on alchemy, the work of Wei Po-Yang, in A.D. 140, called the 'Union of the Three Principles', *Tshan Thung Chhi*. That is a date earlier than alchemy in Europe by about six hundred years.

I might perhaps now give you one or two quotations from Taoist writings, and I would like to do so from the *Tao Tê Ching*, just to show you what is there. One of the queer things about the Taoists is their emphasis on the feminine, reminding us of Goethe's *ewig weibliche*:—

> 'The Valley Spirit never dies.
> It is named the Mysterious Feminine
> And the doorway of the Mysterious Feminine
> Is the base from which Heaven and Earth sprang.
> It is there within us all the while;
> Draw upon it as you will, it never runs dry.'
>
> (ch. 6; tr. Waley.)

This emphasis on the feminine may be regarded as a symbol for the receptive approach to Nature characteristic of the Taoists. The feudal attitude to the organization of society was intensely masculine. The Taoists' attitude in the investigation of Nature was feminine in the sense that the investigator cannot approach Nature with preconceived ideas. 'The Sage is like Heaven and Earth, he covers all things impartially.' The impartial approach without bias, asking questions in a humble way, the spirit of humility in the face of Nature, was understood by the Taoists, as when they speak of the 'valley which receives all the water that flows down into it'. I believe they sensed that the scientist must approach Nature in a spirit of humility and adaptability, and not with that masculine ordering sociological determination which the Confucians had. Here is the interesting passage in which it is said that the highest good of life is like water:—

> 'The Highest Good is like that of Water.
> The goodness of water is that it benefits the myriad creatures;
> Yet itself does not wrangle,
> But is content with the places that all men disdain.
> This is what makes water so near to the Tao.'
>
> (ch. 8; tr. Waley.)

> 'He who knows the Male, yet cleaves to what is Female,
> Becomes like a ravine, receiving all things under Heaven,
> And being such a ravine
> He knows all the time a power he never calls on in vain. . . .

He who knows glory, yet cleaves to ignominy,
Becomes like a valley, receiving all things under Heaven,
And being such a valley
He has all the time a power that suffices. . . .'

(ch. 28; tr. Waley.)

Then, again, there is a fine story in Chuang Tzu which shows what the Taoists meant by 'the Way' or the 'Order of Nature'. His disciples were trying to find out what he meant by the Tao, and said: 'It surely can't be in those broken tiles over there?' He replied: 'Yes, it is in those broken tiles.' The disciples asked a series of such questions, and ended by saying: 'It surely can't be in that piece of dung?' But the reply was: 'Yes, it is everywhere'. That may be interpreted in a religious mystical sense, as referring to the universal operation of a creative force, but the connection of Taoism with the beginnings of science shows, I think, that we should interpret it in a naturalistic way; the idea of the Order of Nature permeating everything.

With this idea in view, you may also notice another story in Chuang Tzu—the famous one about the butcher and the King of Wei. The King, observing his butcher cutting up a bullock for the table, noticed that the man did it with three strokes of his hatchet, so he asked how was he able to do that. The butcher answered: 'Because I have been studying all my life the Tao of the bullock. I who have studied the Tao of the animal can do it in three strokes and my hatchet is as good as it was before. Others do it in fifty strokes and then blunt their axes.' Here we have an indication of primitive anatomy, a beginning of the understanding of the nature of things.

In trying to show you the pre-scientific element in Taoist philosophy I have mentioned alchemy and astronomy and referred now to anatomy. That is well established, but what is not so clearly seen is the full nature of the division between the Taoists and the Confucians. I want to go on to emphasize this, because I think it is vital for the understanding of primitive society in China, both pre-feudal and feudal.

In the *Tao Tê Ching* you will find a number of passages which appear to be against knowledge. For example, in the nineteenth chapter:—

> 'Banish wisdom, discard knowledge,
> And the people will be benefitted a hundredfold.
> Banish human kindness, discard morality,
> And the people will be dutiful and compassionate.
> Banish skill, discard profit,
> And thieves and robbers will disappear.
> If when these three things are done
> They find life too plain and unadorned,
> Then let them have accessories,
> Give them Simplicity to look at
> The Uncarved Block to hold.
> Give them Selflessness
> And Fewness of Desires.'

(ch. 19; tr. Waley.)

'Banish wisdom, discard knowledge' surely sounds odd, for the Taoists were among the earliest thinkers.

But we have just the same story at the end of the Middle Ages in Europe. W. Pagel, the historian of science, has demonstrated how in the seventeenth century and the time of Galileo the theologians in the Christian Church were divided into two camps, on the one hand the rationalists and on the other the mystical theologians. They were equally divided about their attitude to the new science which was growing up by the work of men like Galileo. You will remember that the rationalist theologians refused to look through Galileo's telescope, because, they said: 'If we see what is written in Aristotle, there is no point in looking through the telescope. If we see what is not written there, it can't be true.' That was a very Confucian attitude. Galileo corresponded rather to the Taoists, who had an attitude of humility towards Nature and were anxious to observe without pre-conceptions. Now the mystical theologians were in favour of science because they believed that things could happen if people did things with their hands. The mystical theologians were backward in one sense because they believed in magic, but they believed in science, too, for in the early stages magic and science are closely connected.

If I believe that by taking a wax statue of the chairman and sticking pins in it I can cause him evil, I am adopting a belief for which there is no foundation, but I do at any rate believe in the efficacy of manual operations, and science is therefore possible. The rationalist theologians and the Confucians were against using their hands. There has, in fact, always been a close connection between this rationalist anti-empirical attitude and the age-old superiority complex of the administrators, the high-class people who sit and read and write, as against the low-class artisans who do things with their hands. Just because the mystical theologians believed in magic, they helped the beginning of modern science in Europe, while the rationalists hindered it.

It is the same story in ancient China. When the *Tao Tê Ching* says 'banish wisdom', it means Confucian wisdom. When it says 'discard knowledge', it means discard social knowledge, discard scholastic Confucian 'knowledge'. You will find several passages in Chuang Tzu where he says: 'What are all these distinctions between princes and grooms? I will not have my disciples observe such absurd distinctions.' So here we are coming upon a political element. I want to establish my point. Banishing wisdom, discarding knowledge means, in ancient Taoism, the offensive against Confucian ethical rationalism, the knowledge of the counsellors of the feudal princes, and does not mean banishing the knowledge of Nature, because that was just what the Taoists wished to acquire. They did not, of course, know how to do it, because they did not develop the scientific experimental method, but they wanted it.

Thus we come upon a remarkable political factor. Before I speak further about it, I would like to emphasize the previous point once more, because it is interesting for those concerned with the history of ethics and mysticism.

We cannot say that all through history rationalism has been the chief progressive force in society. Sometimes it undoubtedly has, but at other times not so, because in the seventeenth century in Europe, for example, the mystical theologians gave a good deal of aid to the scientists. After all, natural science was then called 'natural magic'. So in ancient China it is quite clear that Confucian ethical rationalism was antagonistic to the development of science, whereas Taoist empirical mysticism was in favour of it. When they spoke about the Tao, 'holding on to the one', etc., you have a stage in which religion is hardly separated from science, because the one may be the One of religious mysticism, or the universal Order of Nature as we understand it in the scientific sense. It probably means both things, and here we stand at the beginnings of both. Fêng Yu-Lan made one of the best remarks on the subject when he said: 'Taoist philosophy is the only system of mysticism which the world has ever seen which is not fundamentally anti-scientific.'

Now let us examine further the political element. We have seen that phrases such as 'Banish wisdom, discard knowledge. . . .' are to be interpreted in the light of 'I do not wish my disciples to understand these absurd distinctions between princes and grooms' —i.e., class-distinctions. The Taoists were against feudal society, but not exactly in favour of something *new*. They were in favour of something *old*, and wanted to go back to the primitive tribal society before feudalism—as they themselves put it, 'before the Great Way decayed' (ch. 18). Before the Great Way decayed, 'before the Great Lie began', there were none of these class-distinctions. One does not have to read far in Chuang Tzu to find how surprisingly outspoken he is. He says, practically in so many words, that the little thief is punished, but the big thief becomes a feudal lord, and the Confucian scholars are quickly flocking around his doors, wanting to become his counsellors! There can be no doubt that the Taoists were enemies of feudal society, and what they wanted, I think, was the primitive tribal society before the differentiation of classes into warriors, lords, and people.

For example, in the passage—'Banish wisdom, discard knowledge'—it says: 'If the people find life too plain and unadorned, give them Simplicity to look at, the Uncarved Block to hold.' These are odd expressions. It occurred to me one day, when thinking about this, that it might mean, not what European translators usually think it means—namely the One of religious mysticism—but the oneness of primitive society before the differentiation of classes. When you get that clue you find some very interesting other clues quickly following. Besides the 'Uncarved Block', the Taoists are often using other symbols of homogeneity, 'the Post', 'the Bag', 'the Log', 'the Bellows', (important in bronze founding), and a word which is translated 'Chaos'. Throughout Taoist thought you have this feeling that society has been spoilt, 'messed about', and that one ought to go back to primitive simplicity, i.e., before the differentiation of classes, before the first feudal lords. 'The greatest carver is he who does the least cutting' (ch. 28).

A very curious thing is to be noticed here. If we read the books containing the most ancient legends of China, like the *Shan Hai Ching*, the *Shu Ching*, the *Tso Chuan*, and the *Kuo Yü*, for example, we find that many of the earliest legendary kings, such as Yao and Shun, are supposed to have fought with men or monsters—it is not quite clear whether animals or men—but the extraordinary fact is that the names of these beings which they fought and destroyed have just the same sort of ring—Huan-Tou, the empty bag; Thao-Wu, the stake or post which has not been carved up. This is a curious coincidence, because it suggests that the beings against whom the first kings fought were really the leaders of that primitive tribal society resisting the first differentiation of classes— great rebels who had to be beaten down. You also get names like San Miao, Chiu Li, etc. (the Three Miao and the Nine Li), which suggest that there may have been confraternities in that primitive society. Moreover, the legends attribute to all these earliest rebels great skill in metal-working. It looks as if the earliest kings or feudal princes recognized bronze metallurgy to be the basis of feudal power over the neolithic peasantry, because of the superior arms which it rendered possible, and therefore they appropriated the technique of metal-working. It looks as if the pre-feudal collectivist society which developed metal-working resisted the transformation into class-differentiated society, and under the legendary labels we should perhaps see the leaders of that society which resisted the change. There is another phrase to be found alongside these curious phrases—'returning to the root'. That has been translated in a religious sense, but I am not sure that it has not a double political meaning, because in the *Shu Ching* (the *History Classic*) you find a phrase 'the root was kept in check and could not put forth shoots' side by side with a remark about the hosts of Kun flying away. Kun was one of the most prominent of these early rebels.

I am now directing attention to the political significance of Taoist philosophy. Throughout the centuries in China there have been secret sects of various kind, adepts of peasant type—secret societies, of course—and even now today, in China, secret societies are still important. All through Chinese history it is always jokingly said: 'Confucianism is the doctrine of the scholar when in office, and Taoism is the attitude of the scholar when out of office', because the scholars have always been in and out of office, in the mandarinate, the Civil Service. Even though in later centuries it became an organized religion with elaborate liturgical worship, Taoism has always been connected with movements against the Government, and in all dynasties—Thang, Sung, Ming —it has been of political significance. I want to draw particular attention to this because it is a thing which is very little appreciated by many who study Taoism in Western Europe.

A book such as the *Tao Tê Ching*, on account of the laconic and lapidary style of ancient Chinese, is susceptible of many interpretations. Western scholars, perhaps following the classical commentators such as Wang Pi, have always adopted the mystical interpretation, but it is interesting to see what a modern Chinese scholar, aware of the political interpretation, makes of a passage. Here I give chap. 11, translated first on the mystical theory and then on the political theory:—

(*a*) 'Thirty spokes together make one wheel
And they fit into nothing at the centre;
Herein lies the usefulness of a carriage.
Clay is moulded to make a pot
And the clay fits round nothing;
Herein lies the usefulness of the pot.
Doors and windows are pierced in the walls of a house
And they fit round nothing;
Herein lies the usefulness of a house.

Thus while it must be taken as advantageous to have something there,
It must also be taken as useful to have nothing there.'

(tr. Hughes; Waley's translation is very similar.)

(b) 'Thirty spokes combine to make a wheel,
When there was no private property
Carts were made for use.
Clay is formed to make vessels,
When there was no private property
Pots were made for use.
Windows and doors go to make houses,
When there was no private property
Houses were made for use.
Thus having private property may lead to profit,
But not having it leads to use.'

(tr. Hou Wai-Lu.)

All this has a clear connection with the interest of the earlier Taoists in natural science, because, as many scholars such as Diels and Farrington have shown in examining Western European antiquity—among the Greeks, for example—there is a distinct connection between interest in the natural sciences and the democratic attitude and relationship, particularly with regard to the power of the merchants. Thus there was a connection between Ionian natural science and commerce in the Eastern part of the Mediterranean basin. It looks as if interest in natural phenomena, natural science, does not flower, does not come to anything, under despotisms or certain kinds of bureaucratisms. I shall return to this point at the end of the lecture.

There is more to be said about the ancient feudal age in China. We have mentioned already that there was a continuity between the Bronze Age in China and the Bronze Age in Europe. Weapons and utensils have similar designs in China, and in the Hallstatt and La Tène cultures in Europe. Now, the analogy is usually made between feudal China and our own European medieval feudal period. But it is very mysterious why feudalism began in Europe, as most people would say, about the third century A.D. and closed at the time of the rise of capitalism, the Renaissance and the Reformation in the fifteenth century A.D.; whereas in China feudalism is so much earlier, from the fourteenth to the second century B.C. The fact is that the analogy between Chinese feudalism and Western European feudalism is not sound. It ought not to be likened with high medieval feudalism, but rather with the society of pre-Roman Europe.

Ancient Chinese feudalism is, I think, analogous with the state of society in the European Bronze Age, or when the Bronze Age was giving place to the Iron Age—about 300 B.C., before the Roman conquest of Gaul. That kind of society is called by archaeologists quasi-feudal society. The essence of it is a series of chiefs, with maybe a High King—something like Conachur of Ireland—and then a series of chiefs in descending ranks, a kind of hierarchy, each one having men-at-arms who are pledged to come and rally round the leaders in case of war. The armies which the Gauls brought together to oppose the Romans were formed of such quasi-feudal levies. Large-scale slavery was not involved. We might thus say that feudalism in Europe lasted from about 1000 B.C., as in China, until the fifteenth century A.D., but that it was overlaid by two or three centuries of City-State imperialism in the shape of the Roman Empire.

Now, it is most significant that the institution of large-scale slavery was not known in ancient China. There is a certain amount of controversy about this, but the balance of evidence seems to be that slavery as understood in the Mediterranean civilizations—Egypt, Babylonia, Rome or Greece—was not known. That is an important fact. Chinese society has always been modelled on a basis not of slavery, but of free farmers, and that has a very important bearing on the humanitarian character of Chinese philosophy in all forms, whether Confucian or Taoist. It is not at all obvious at first sight what was the reason for this, because there would have been nothing to prevent the ancient Chinese from having a large slave population derived from captives taken in war, people of the Mongol or Hunnish tribes to the North or the Tibetans and Tanguts in the West.

It is an important question, and brings us back to the question of ethics. It can, of course, be said that it was not in accordance with

Confucian ethics. That kind of explanation is not very satisfying, however. We want to look for something more concrete. Philosophy in general cannot be studied apart from the actual concrete social background, including many technological factors. Following one of the greatest experts on the Chinese Bronze Age, Creel, I would like to suggest the importance of the relation between the technological military level of the ruling class in relation to the people. Take the extreme case of the medieval knight in Western Europe, with his steel armour from top to toe, his lance and his sword, mounted on his horse, also armoured. He was able to ride into a mass of peasants and mow them all down without their being able to defend themselves. It is a commonplace—we learn it at school—that it was the coming of gunpowder to Europe (a Chinese discovery, incidentally) which broke up the feudal power by removing the technical superiority in arms of the knightly class.

What was the situation, then, in ancient China? There the crossbow—a most powerful weapon—was invented centuries before anywhere else. We know that the men of the feudal levies in ancient China (by that I mean from 800 to 300 B.C.) were armed with powerful bows. But at the same time protective armour was very little developed. The archaeologist Laufer has written a fine monograph on Chinese armour. It arises very late, and in early times you only get protective clothing made of bamboo and wood. Moreover, there are in the *Tso Chuan* countless stories of feudal lords being killed by arrow-shots. If the mass of the people as a whole were in possession of a powerful offensive weapon, and the ruling class were not in possession of superior defensive means, one can see that the balance of power in society was different from what it was in, for example, the time of the early Roman Empire, where the disciplined legions were rather well armoured, with bronze and iron. A slave population was possible because it was not in possession of the arms and armour of the legionaries, nor did it have access to powerful bows. The principal Roman weapons were always the spear and the short sword. We know what trouble the slaves could give on the few occasions in which they did gain access to substantial stores of weapons, as in the revolt of Spartacus. In China it was a different story, because from an early date the people had crossbows and the lords had poor defensive armour. If that was the case, it means that the people in China had to be persuaded, rather than cowed by force of arms, and hence the importance of the Confucians. In the fourth century B.C., in a State such as Sung or Wu or Chhu, the people on whom the lord depended might well desert to his opponent suddenly on the field of battle. They had to be convinced of the justice of their cause. To effect that it was necessary to have a class of 'sophists' which afterwards became the Confucians, to commend

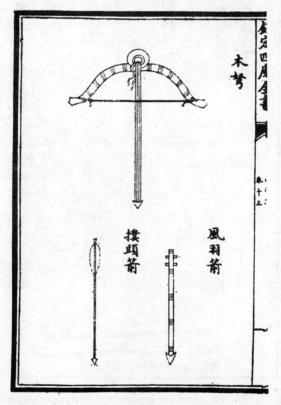

FIGURE 22 The Chinese crossbow and bolts, drawings from the *Wu Ching Tsung Yao* (Compendium of Important Military Techniques), by Tsêng Kung-Liang, A.D. 1044.

to the mass of the people the activities and virtues of the feudal lord. and to gather them around him.

If that was the case we can understand much better the humanitarian and democratic character of the Confucian philosophers. Mêng Tzu was one of the first thinkers in history to defend the

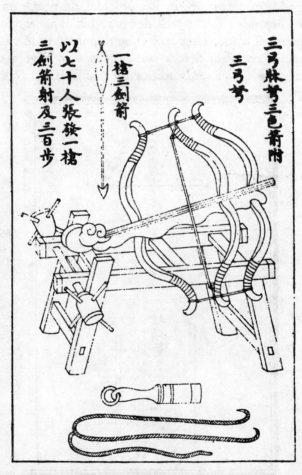

FIGURE 23 The Chinese crossbow applied at artillery scale as an arcuballista; another drawing from the *Wu Ching Tsung Yao* (A.D. 1044). Three springs are combined on a frame and armed by a winch and handspikes. Three quarrells are shot off simultaneously to a range of 300 double-paces (c. 600 yards).

right of the people to overthrow and kill tyrants. The aversion from the appeal to force—a very specific character of Chinese society—may be connected with these facts. There was no slavery, apart from certain kinds of domestic slavery; no mass slavery, such as one found in the Mediterranean civilizations; the mass of the people who carried about the stones for the monuments of Egypt and Babylonia or worked in the Spanish mines described by Diodorus Siculus, or manned, in late Roman times, the *latifundia*. And since there was no slavery in China, may we not draw a connection between that and the technological significance of China for the outside world?

The famous German archaeologist Diels and many other historians of science have suggested that the failure of applied science to develop in early Mediterranean civilization was due to the fact that slavery existed, and hence there was no labour problem and no object in inventing labour-saving devices. This is a commonplace point.

Now if this was not quite the case in China, there may be a connection between social status and the technologically advanced position of China in those periods. Europeans of today are under the domination of the ideas of the last century, and do not realize that, if you go back three or four hundred years, China was a better place to live in than Europe. In Marco Polo's time Hangchow was like a Paradise compared to Venice or the other dirty towns of Europe. Early travellers like John of Monte Corvino have the same story to tell. The standard of life was higher in China than in Europe in those days.

The inventions of gunpowder, paper, printing, and the magnetic compass are generally acknowledged—I think correctly—to have been transmitted from China to Western Europe. There are many other inventions of the same kind, which are more unfamiliar. I propose to describe one of the most important of them now.

The history of animal harness is of extreme importance in connection with the history of social institutions, because if you have slavery you do not need an efficient harness for animals. If you have an efficient harness for animals, you can do without

視聽自民圖

FIGURE 24   A late Chhing illustration of the famous statement in the *Shu Ching* (Historical Classic), *c.* 9th century B.C., 'Heaven sees as the people see, Heaven hears as the people hear' (Kao Yao Mo ch., from *Shu Ching Thu Shuo*). Some are quarrelling, others leading a malefactor along; all such social problems need the decisions of rulers. If they rule with justice and righteousness, Heaven's approval is manifested by the people; if otherwise, they are overthrown and their place taken by another dynasty.

slavery. If the Egyptians had had efficient animal harness, they might have used animals for transporting the vast blocks of stone for building the Pyramids. They did not have it. We know from the carvings, which may be seen in dozens in the British Museum, that they used men for carrying out this muscular labour.

The story is this. For four thousand years—from 3000 B.C. in the earliest Sumerian pictures, down to A.D. 1000 in Europe—the only harness known was what we may call the 'throat-and-girth' harness, where the pull of the chariot was taken by the yoke at the point where the belly-strap joins the throat-strap. This harness is exceedingly inefficient because the animal fitted with it cannot pull more than 500 kg. The reason is obvious, because the main pull comes on the throat, and the horse tends to be suffocated (cf. Fig. 10 on p. 88).

Modern harness, on the other hand, as we know it, is different. Modern harness is the 'collar' harness, whereby the animal is enabled to exert the whole of its force, since the collar pulls on the shoulders. It is hardly believable that the ancient harness continued up to A.D. 1000 in Europe. I must mention at this point how these facts were investigated. An extremely ingenious retired French officer, Lefebvre des Noëttes, an adept at asking simple questions which nobody could answer, inquired if there was anyone who could inform him when the modern collar harness originated. Nobody had any idea, so he proceeded to look at all the carvings of animals in the museums from all civilizations, and at the illustrated manuscripts in the libraries. From the earliest Sumerian and Babylonian civilizations at one end up to A.D. 1000 —the early Middle Ages—as we have said, the inefficient 'throat-and-girth' harness was used, while after that time in Europe the 'collar' harness came into use. But there was one exception— China. In China they had what I may call the 'breast-strap' harness. A trace on each side of the animal is held up by straps, and the pull comes on the shoulder. The Chinese chariot did not have, like the Roman or the Greek chariot, a straight pole or shaft, but a curved one attached to the breast-strap half-way along its length. We may also call this the 'postillion' harness, for it is still used in the south of France today, and called 'attelage de postillion'. The pull comes in the right place. The animal is not stifled, and can pull a heavy load. Thus in the Han bas-reliefs you will find that the Chinese chariots were three or four times larger than anything in Europe. Instead of having two men—a charioteer and the lord—standing, or the single Babylonian or Greek warrior, you have a whole bus, about four or five or even seven people sitting in the chariot, and even a roof—one of those large, curving roofs, on the chariot. It is a totally different matter from the Western chariot. Now it is clear that the connection between the collar harness and the 'breast-strap' harness is rather close, for if you imagine the collar to be flexible instead of rigid, it would take up the position of the postillion or 'breast-strap' harness when the pull came on it.

What of the dates? The 'breast-strap' harness goes back at least to 200 B.C.—the beginning of the Han dynasty in China—and all through Chinese history after the feudal period you get it. Moreover, at the time when the 'collar' harness first appears in Europe, the 'breast-strap' harness had been there for only two hundred years. The other essential fact is that about the end of the fifth century A.D. you find on Buddhist cave-paintings and sculptures in N.W. China both the 'collar' harness and the 'breast-strap' harness; which seems a rather clear indication that the efficient harness came to Europe between A.D. 600 and 1000. Those who think that everything good has come out of Europe, and that the 'Great White Race' are the most wonderful people on earth, and that wisdom was born with them, should study a little history to realize that a great many of the things on which Europe prides itself were not originally in Europe at all. I think it is clear that the efficient animal harness is one of these things. What were the social conditions which led to its coming to Europe is another matter; it may have been the building of the cathedrals, where the necessity arose once more for carrying heavy blocks of stone. By that time ancient Mediterranean slavery had died out and the feudal age had come, because feudal society was a great deal stronger than the society of the decaying Roman Empire, with its latifundia (the great estates). Since slavery no longer existed in

Europe, it was necessary to have an efficient animal harness, and the place to get it from was that part of the world where there had never been slavery—namely China.

It is to be hoped that this lecture has not been unsatisfying to the audience. I have not had any special thesis to bring before you. I have simply tried to sketch a certain pattern of society—Chinese feudal society—and mention its relations to Western European society; out of which I thought would come, and I believe have come, a number of points of interest to anyone who is thinking about such questions as ethics, rationalism, and culture as a whole. We have seen that rationalism is not always the most progressive force in society. We have seen that the status of military technology may deeply affect the crystallization of social philosophy. We have seen that a moral question such as slavery may be closely connected with technical factors. Philosophical and ethical thought can never be dissociated from their material basis.

If I am to say one word in conclusion about the further problems, the wider problems, of the rise of modern science and technology, I might end as follows. There is no time to justify it, but I believe that, in spite of the excellence of ancient Chinese philosophy and the importance of the technological discoveries made by the Chinese throughout later history, Chinese civilization was basically inhibited from giving rise to modern science and technology, because the society which grew up in China after the feudal period was unsuitable for these developments. When European feudalism decayed about the sixteenth century, capitalism took its place. There was the rise of the merchants to power, bringing first mercantile and then industrial capitalism. But in China, when Bronze Age feudalism decayed and the Imperial Age came, there was no question of a temporary suspension of feudalism by an imperialist City-State like Rome. Something quite different happened. Ancient feudalism in China was replaced by a special form of society to which we have no parallel in the West. This has been called Asiatic bureaucratism, in which all the lords have been swept away except one—the Son of Heaven, the Emperor, who rules the country and collects all the taxes through a gigantic bureaucracy. The people who made that bureaucracy, the mandarinate, were the Confucians, and for two thousand years the Taoists fought a collectivist holding action, only to be justified by the coming of socialism in our own time. All this was something unknown in the West, and requires special and intense study, but it certainly had one big effect—to prevent the rise of the merchant class to power. To ask why modern science and technology developed in our society and not in China is the same thing as to ask why capitalism did not arise in China, why was there no Renaissance, no Reformation, none of those epoch-making phenomena of that great transition period of the fifteenth to the eighteenth centuries.

That was what I wanted to explain here. I should like to end by saying that I would very much recommend to anyone the experience of having a closer look at the great classics of Chinese philosophy, as well as the parallel course of technology in China. It is so exciting because Chinese culture is really the only other great body of thought of equal complexity and depth to our own —at least equal, perhaps more, but certainly of equal complexity; because, after all, the Indian civilization, interesting though it is, is much more a part of ourselves. Our language is Indo-European, derived from Sanskrit. Our theology embodies Indian asceticism; Zeus Pater derives from Dyaus Pithar. There is much more in common between Indian and European civilization, just as there is in the visible type. I often used to think when walking about the streets of Calcutta, that if the pigment was taken out of the skin of many of the people, their features would be quite similar to those of our immediate friends and relations in England. But Chinese civilization has the overpowering beauty of the wholly other, and only the wholly other can inspire the deepest love and the profoundest desire to learn.

5

THOUGHTS ON THE SOCIAL RELATIONS OF SCIENCE AND TECHNOLOGY IN CHINA

First published in *Centaurus*, 1953, **3**, 40.

One of the most fascinating questions in the comparative history of science, as it might be called, concerns the failure of the two great Asian civilizations, China and India, to develop spontaneously *modern* science and technology. It is unfortunate that their contributions to ancient and medieval science are not better appreciated, since only with that background in mind can the unique appearance of mathematized natural philosophy in Europe be comprehended. Before the fourteenth century A.D., Europe was almost wholly receiving from Asia rather than giving, especially in the field of technology. What can be said about the social milieu which produced that accomplishment and that failure?

There seems no doubt that in early periods there was feudalism in China. It might perhaps be described as a 'Bronze Age' proto-feudalism. It covered the period, roughly speaking, from the middle of the second millennium B.C. down to about 220 B.C., at which time the first unification of the Empire took place. But from then onward, the use of the word 'feudalism' seems more and more difficult because whereas the earliest period bears some resemblance to European medieval feudalism, the later periods are very different. The social system which emerged has been called Asiatic bureaucratism, or, as some Chinese scholars prefer,

FIGURE 25 A late Chhing illustration of a sentence in the *Shu Ching* (Historical Classic), *c.* 9th century B.C.; 'In a single day ten thousand details, the germs of things to come, arise; (the good ruler must attend to them all)'. The legendary emperor Shun is shown going laboriously through the files brought to a terrace by assiduous secretaries, (Kao Yao Mo ch., from *Shu Ching Thu Shuo*).

bureaucratic feudalism. In other words, the ending of the first feudalism in China did not give rise to mercantile capitalism and industrial capitalism, but brought about instead a bureaucratic system involving the loss of the aristocratic and hereditary principle from Chinese society. What happened, one might almost say, was that when the individual feudal lords of the intermediate levels ceased to exist, there remained only one great feudal lord, namely the emperor, governing and collecting taxes through a gigantic bureaucracy.

The members of this bureaucracy did not fully form a hereditary group, and so did not constitute a class in the customary sense of the word when used in relation to European societies. It was, as it were, an estate, and it had fluidity; families rose into it and sank out of it. It was in fact a learned élite. As is well known, at a later period entry was through the State examinations, a system which began during the Han dynasty, in the first or second century B.C., but did not attain its real flowering until the Thang dynasty in the seventh century; and then it went on until the coming of the Republic in 1912. The examinations—again this is very generally known—were entirely based on literary and cultural subjects, and did not include subjects which could, in any sense, be called scientific,[1] but still, the examinations were quite difficult; indeed, when the extreme complexity of the Chinese language and literature is borne in mind, very difficult. But there were also, at different times and periods and in varying degrees, ways of getting around the examinations and entering the civil service hierarchy without passing through them. There was the 'Yin privilege', by which the sons of bureaucrats were given an easier entry than those who came from outside. But on the whole, as far as individuals were concerned, the class was fluid. Families were rising into it and sinking out of it all through the centuries. It is also known that at some periods, the possibilities for a man of rural peasant family were quite considerable, and it was sometimes the custom for farmers to club together to pay for a tutor for some particularly promising young man in order that he might enter

[1] With occasional exceptions, as under Wang An-Shih in the Sung.

FIGURE 26 A late Chhing illustration of a legend relating to Yü the Great, the hydraulic engineer culture-hero of Chinese antiquity, given in the *Shu Ching* (Historical Classic) *c.* 9th century B.C.; 'When I started the work (of regulating the waters) I married at Thu-shan, (but afterwards I went back only for four days there, and hurried away to resume my charge).' His wife is holding their son Chhi, (I Chi ch., from *Shu Ching Thu Shuo*).

the Imperial service; investing, as it were, for benefits that would then accrue to his native place.

If one investigates the origins of this bureaucratic system which impressed its character so deeply on Chinese society, one comes upon several factors, geographical, hydrological and economic. The reason which has been given by one of the earliest occidental economic historians of China, K. A. Wittfogel, for the origin of the bureaucracy was that it was conditioned by the immense and early growth of hydraulic engineering works in Chinese society. I found when I was in China that that viewpoint is quite widely accepted by Chinese scholars who put, however, a somewhat different emphasis on it. The effect of the importance of irrigation and water-conservation works in Chinese history is indeed undoubted. Probably no country in the world has so many legends about heroic engineers, for example the legendary emperor Yü the Great, who 'controlled the waters' for the first time in Chinese history. The rainfall in China is of course extremely seasonal, because it is a monsoon area, and also highly variable from year to year. When you consider how necessary irrigation was for wet rice cultivation in the centre and south and for the cultivation of the loess lands in the north, and when you add the constant flood danger requiring water-conservancy techniques, you see at once how extremely important these works were. We know that they started already in the feudal period (fifth century B.C.). There is moreover a third reason why the water control system of the country was profoundly important, and that was because it provided a means of transportation. Since taxes were collected, or military supplies brought together, in the form of kind and not money, the accumulation of rice and other grain at the capital required a method of heavy transport as by barges on canals. So there were three needs—irrigation, water-conservancy, and tax-grain transportation—which required a water economy to come into existence. Whereas Western scholars have suggested that the origin of the 'mandarinate' could be traced to the fact that control had to be exercised over the millions of men who were brought together to carry out these works, many

Chinese scholars whom I have read and listened to consider that the deeper reason why this domination of the society by a 'civil service' took place was because there was always the tendency to transfer control to central authority—in other words that the carrying out of water-work plans tended to transcend the boundaries of the estates of feudal lords. As a matter of fact this is stated, in so many words, in one of the great Chinese books, the *Yen Thieh Lun* (Discourses on Salt and Iron) written in 81 B.C.

This remarkable work, which reads like the records of a party conference (I should say a Conservative Party conference) is, in fact, the dramatised account of one which actually took place about the nationalization of the salt and iron industry, recommended as early as 400 B.C. and actually put into operation in 119 B.C. The Lord Chancellor opens one of the speeches by saying that we all realize that small local lords or governors are responsible for small amounts of territory, but the development of rivers, canals, and sluices must devolve upon the central authority. He was stating what was to remain a permanent feature of Chinese society. One of the earliest efforts of the mandarinate was in fact the nationalization of salt and iron in the former Han dynasty. These were the most important, perhaps the only, things that travelled from town to town. Everything else could be made *in situ*, whether in weaving or the preparation of food, on the farm or in the local town, but salt and iron radiated from proto-industrial centres, salt from the sea coasts or brine fields, iron from the places where ore was found, and these were therefore the two commodities most suitable for control and 'nationalization'. The interesting thing about the arguments which were put forward is that both the Confucian scholars who were criticizing the Han bureaucrats, and the bureaucrats themselves, were violently against the merchants. There is, in fact, quite a mass of interesting evidence about the growth of a merchant community at the time when the first Chhin Emperor unified the country and started the first centralized dynasty (220 B.C.). There is a special chapter in the *Shih Chi*[1] about the merchants of that time. Some were extremely

[1] Ssuma Chhien's 'Historical Memoirs', written about 90 B.C.

wealthy; some were ironmasters, others were concerned with salt. Their power was immediately attacked by the early bureaucrats and rapidly destroyed. Sumptuary laws were enacted against them, and severe monetary taxes inflicted on them.

There is probably no other culture in the world where the conception of the civil service has become so deeply rooted. I myself had no idea of it when I first went to China, but you can find it everywhere there, even in the folk-lore. Instead of stories about heroes and heroines becoming kings or princesses, as in Europe, in China it is always a matter of taking a high place in the examinations and rising in the bureaucracy, or marrying an important official. This was, of course, the only way in which to acquire wealth. There is a famous saying (current till recently) that in order to accumulate wealth you must enter the civil service and rise to high rank (*Ta kuan fa tshai*). The accumulation of wealth by the bureaucracy was the basis of the phenomenon often described by Western people in China as 'graft', 'squeeze', and so on, and of which so many complained. The attitude of Westerners, however, has been prejudiced by the fact that in Europe religion and moral uprightness had a historical connection with that quantitative book-keeping and capitalism which had no counterpart in China. At no time in Chinese history were the members of the mandarinate paid a proper salary, as we should think natural in the West. There were constant efforts to do so, decrees were always being issued, but in point of fact it was never done, and the reason is probably because the Chinese never had a full money economy.

Taxes had to be paid in kind, and transmitted in kind to the central authority, using the methods of fluviatile navigation to which I have already referred. It became inevitable that this tribute should be 'taxed at the source' (from the point of view of the emperor), and there are many expressions in Chinese for this state of affairs, one of the best being *chung pao* ('middle satisfied'), the point being that the peasants were not satisfied because of having to pay more than they thought they ought, and the emperor was not satisfied either, but the officials in the middle were quite satisfied because they were 'taking a cut off the joint' at every stage. A special word with no moral connotations is needed for this phenomenon, to indicate that it was a natural feature of Chinese medieval society. When the bureaucrat, whether a magistrate of a city, or the governor of a province, or a *Chuang Yuan* with eight cities under his charge, had accumulated his capital, what he did with it, apart from expenditure on luxuries (and this would be quite natural in any large official family) was invariably to invest it in land. Land purchase was the only method of investment, and the result was a gradual increase in the number of tenant farmers. Before the overthrow of the Kuomintang, forty or fifty per cent of the peasants were tenants, and most of their farms were uneconomically small.

I will now turn to another aspect of bureaucratic influence, which was always exerted against the merchants. The despising of the merchant was a very old characteristic in Chinese thought (and much in contrast with Arabic ideas); in the classical enumeration of the four ranks of society, the scholars came first, then the farmers, third the artisans and fourth the merchants; the merchants were supposed to be socially the lowest (*Shih, Nung, Kung, Shang*). There was of course in China nothing resembling the caste system, or even a class system in the orthodox sense of the word, but still, as a stratum of society, the merchants were certainly supposed to be the least socially respectable. It is nevertheless true that the merchants in China ultimately formed themselves into guilds, but one has to take a closer look at what they were like. I know something of them, because I have stayed in large houses belonging to merchant guilds. For instance, the University of Amoy set up its library during the war at Changting in a large house of many courtyards, which was the guild-house formerly used by the Chiangsi merchants who came to trade in Fukien. There is no question that there were guilds, but as several useful books have described them, they were different in many ways from the merchant guilds in Europe. They were more like mutual benefit societies, insurance organizations, protecting against loss occasioned in transit, and the like, but the one thing they never did was to acquire real control or power in the cities where the merchants lived and carried on their trades, or organized their small production workshops.

There was thus an essential difference between the guilds in China and those in the west, just as much indeed as there was between the city in China and the city in the west. Perhaps it can be summed up by saying that the conception of the city-state was unknown in Chinese culture and civilization and the cultures that derived from it.[1] You have to set against the European conception of the city-state the Chinese conception of a city with its walls, surrounded by many villages from which the people come for the sake of market and trade, and with the headquarters of the magistrate or provincial governor appointed from the Imperial Court, responsible to no one except his superior officials in the bureaucratic hierarchy. There would also be a military mandarin, and the two would have their offices in the town. It would, in a sense, be a walled, fortified city 'held for the Crown' by the responsible local officials. There is nothing in Chinese history resembling the conception of a mayor or burgomaster, aldermen, councillors, masters and journeymen of guilds, or any of those civic individuals who played such a large part in the development of city institutions in the west. These things were quite unknown. A phrase comes to one's mind regarding the cities in the West: *Stadtluft macht frei*—(A man can become free by entering the city and getting permission to live and work there). That is inconceivable in Chinese society. Another germane phrase would be *bürgerliche Rechtssicherheit*—(Security under the laws of the boroughs)—the European merchants freely associating in their towns, and winning charters and advantages of all kinds from the feudal society which environed them. That is all foreign to Chinese culture and thought. Sir John Pratt has brought it out when he relates how the merchants in Shanghai about 1880 appealed to the Imperial Government in China for some kind of state charter which would permit them to elect a mayor or burgomaster, aldermen and so on, in fact to set up

[1] The city-state conception may perhaps be applicable to certain small States in Central Asia, however (W. Eberhard).

all the institutions associated with a city in the West. One can imagine the mystification produced at the Imperial Court at Peking
when the request arrived there. Such lack of understanding was
characteristic of both sides at the time.

There cannot be much doubt (as we can now see) that the failure
of the rise of the merchant class to power in the State lies at the
basis of the inhibition of the rise of modern science in Chinese
society. What the exact connection was between early modern
science and the merchants is of course a point not yet fully elucidated. Not all the sciences seem to have the same direct connection with mercantile activity. For instance, astronomy had
been brought to quite a high level in China. It was an 'orthodox'
science there because the regulation of the calendar was a matter
of intense interest to the ruling authority. From ancient times the
acceptance of the calendar promulgated by the Emperor had been
a symbol of submission to him. On account of a great sensitivity
to the 'prognosticatory' aspect of natural phenomena, the Chinese
had amassed long series of observations on things which had not
been studied at all in the West, for example auroras. Records of
sun-spots had been kept by the Chinese, who must have observed
them through thin slices of jade or some similar translucent
material, long before their very existence was suspected in the
West. It was the same with eclipses, which were supposed to have
a fortunate or antagonistic effect on dynastic events.

Then there were the 'unorthodox' sciences, for example
alchemy and chemistry, which were always associated with Taoism. Neither astronomy nor chemistry could enter the modern
phase, however, in the Chinese environment.

In the West the merchants seem to have been connected especially
with physics, a science which in China had always been particularly
backward except for the brilliant practical development of the magnetic compass. Perhaps this was due to the need of the merchants for
exact measurements. The merchant could hardly carry on his trade
without them. He had to take a lively interest in the actual properties of the things with which he was concerned. He had to know
what sort of weight they were, what they were good for, what sort

of lengths or sizes they came in, what containers would be necessary, and so on. Along such lines as that one might look for the connection of a mercantile civilization with the exact sciences. But besides the merchandise there was also the transport. Everything which had to do with nautical construction and efficiency was of interest, and had always been of interest, to the merchants of Europe's city-states.

If this is the case, it is precisely in the inhibition of the rise of merchants to power in the state that we have to look for the reasons for the inhibition of modern science and technology in Chinese culture. Another aspect of the matter is the old question of the antagonism between manual and mental work which has run through all ages and all civilizations. To Greek *theoria* and *praxis* correspond Chinese *hsüeh* and *shu*. It seems that no one can fully overcome that tradition, no one can advance to the point at which there is equal participation of hand and brain, so absolutely essential in scientific work, no one can succeed in bringing them together, except the merchant class when it succeeds in imposing its mentality on the surrounding society. That was simply never possible in China. There was a restriction of technology to an eotechnic level—seen for instance in the use of wood for gears instead of metal.

Yet here we have one of the most extraordinary paradoxes in history. Few people as yet realize what an enormous technological debt Europe owes to China during the first fourteen centuries of our era. While the old Chinese bureaucratic society was certainly inferior to the society of the European Renaissance in technical creativity, it had been much *more* successful than European feudalism, or the Hellenistic slave-owning society which had preceded it. China contributed things like the efficient horse harness, the drawloom, the sternpost rudder, the first cybernetic machine, the earliest type of vaccination, and even so simple a device as the wheelbarrow—all these (when they travelled) came across from east to west, and not *vice versa*. The strangest paradox is that the very people who by the nature of their society, if I am right in this diagnosis, were prevented from developing, as Europe did at the beginning of the Renaissance and the rise of capitalism, a state of society in which iron would become the basis of the first world-uniting civilization, had, in fact, mastered the difficult art of iron-casting thirteen centuries before the West. We know that cast iron was very uncommon there before the fourteenth century A.D. It may have been produced occasionally by the Romans, but was certainly employed on a widespread scale by the Chinese in the first century B.C. It was in fact an ancient art in China, and the same is to be said also for the iron plough-share, and not only the plough-share, which travelled from east to west, but the mould-board as well. The Chinese were the first to introduce the mould-board—all this in a society unable to advance to the high metallurgical level of the later European societies.

If one asks who was the first to appreciate this difference between Asian and Western society, the answer might well be François Bernier, a French traveller who was physician to Aurangzeb, one of the last of the Mogul emperors. In his book he has some most remarkable pages. I was fortunate enough to get a copy of it in Calcutta and I shall always remember the excitement with which I read it. Written about 1670, in it is raised the question 'whether it is an advantage or disadvantage to the State if the King is the owner of all the land, and not to have the *meum* and *tuum* which exists among ourselves'. He came to the conclusion that it was a 'disadvantage' for a country to have that type of society which we call Asian bureaucratism He has a lot to say about the position of what corresponded to the mandarinate. In India it was not exactly a mandarinate, but still a civil service system, a non-hereditary bureaucracy to which appointments were made by the Mogul emperors.

In concluding, one might suggest that Asian bureaucratism is by no means characteristic only of East Asia. There remains the great problem of Islamic science and society. As is well known, Arabic science was for 400 years much ahead of European. Now it would seem that the earlier Islamic society was really very mercantile. The Prophet himself has many words of praise for the merchants, but few for agriculturalists, and one might consider the Arab towns and cities on the edge of the deserts to be of a mercantile character, the desert taking the place of the sea. When the conquests took place, however, and the Caliphate was established in Baghdad, there came a movement to organize the mechanism of government more fully and introduce a much more bureaucratic state, similar to that which had existed in earlier times in Persia, and was nearer to the Chinese system. So perhaps what began in Islamic civilization as a mercantile culture, ended by being thoroughly bureaucratic, and to this might possibly be ascribed the decline of Arabic society and particularly of the sciences and technology. But all that, of course, would be another story.

6

SCIENCE AND SOCIETY
IN EAST AND WEST

First published in Bernal Presentation Volume (London, 1964),
and in *Science and Society* 1964, **28**, 385 and *Centaurus*, 1964, **10**, 174.

When I first formed the idea, about 1938, of writing a systematic, objective and authoritative treatise on the history of science, scientific thought and technology in the Chinese culture-area, I regarded the essential problem as that of why modern science had not developed in Chinese civilization (or Indian) but only in Europe? As the years went by, and as I began to find out something at last about Chinese science and society, I came to realize that there is a second question at least equally important, namely, why, between the first century B.C. and the fifteenth century A.D., Chinese civilization was much *more* efficient than occidental in applying human natural knowledge to practical human needs?

The answer to all such questions lies, I now believe, primarily in the social, intellectual and economic structures of the different civilizations. The comparison between China and Europe is particularly instructive, almost a test-bench experiment one might say, because the complicating factor of climatic conditions does not enter in. Broadly speaking, the climate of the Chinese culture-area is similar to that of the European. It is not possible for anyone to say (as has been maintained in the Indian case) that the environment of an exceptionally hot climate inhibited the rise of modern natural science.[1] Although the natural, geographical and climatic settings of the different civilizations undoubtedly played a great part in the development of their specific characteristics, I am not inclined to regard this suggestion as valid for Indian culture. The point is that it cannot even be asserted of China.

From the beginning I was deeply sceptical of the validity of any of those 'physical-anthropological' or 'racial-spiritual' factors which have satisfied a good many people. Everything I have experienced during the past thirty years, since I first came into close personal contact with Chinese friends and colleagues, has only confirmed me in this scepticism. They proved to be entirely, as Andrea Corsalis wrote home so many centuries ago, 'di nostra qualità.' I believe that the vast historical differences between the cultures can be explained by sociological studies, and that some day they will be. The further I penetrate into the detailed history of the achievements of Chinese science and technology before the time when, like all other ethnic cultural rivers, they flowed into the sea of modern science, the more convinced I become that the cause for the break-through occurring only in Europe was connected with the special social, intellectual and economic conditions prevailing there at the Renaissance, and can never be explained by any deficiencies either of the Chinese mind or of the Chinese intellectual and philosophical tradition. In many ways this was much more congruent with modern science than was the world-outlook of Christendom. Such a point of view may or may not be a Marxist one—for me it is based on personal experience of life and study.

For the purposes of the historian of science, therefore, we have to be on the watch for some essential differences between the aristocratic military feudalism of Europe, out of the womb of which mercantile and then industrial capitalism, together with the Renaissance and the Reformation, could be born; and those other kinds of feudalism (if that was really what it was) which were characteristic of medieval Asia. From the point of view of the history of

[1] Cf. the writings of E. Huntington, for instance, *Mainsprings of Civilization* (New York, 1945).

science we must have something at any rate sufficiently different from what existed in Europe to help us solve our problem. This is why I have never been sympathetic to that other trend in Marxist thinking which has sought for a rigid and unitary formula of the stages of social development which all civilizations 'must have passed through'.

Primitive communalism, the earliest of these, is a concept which has evoked much debate. Though such a phase is commonly rejected by the majority of Western anthropologists and archaeologists (with, of course, some notable exceptions such as V. Gordon Childe), it has always seemed to me eminently sensible to believe in a state of society before the differentiation of social classes, and in my studies of ancient Chinese society I have found it appearing through the mists clearly enough time after time. Nor at the other end of the story is there any essential difficulty in the transition from feudalism to capitalism, though of course this was enormously complex in detail, and much has still to be worked out. In particular the exact connections between the social and economic changes and the rise of modern science, that is to say, the successful application of mathematical hypotheses to the systematic experimental investigation of natural phenomena, remain elusive. All historians, no matter what their theoretical inclinations and prejudices, are necessarily constrained to admit that the rise of modern science occurred *pari passu* with the Renaissance, the Reformation and the rise of capitalism.[1] It is the intimate connections between the social and economic changes on the one hand and the success of the 'new, or experimental' science on the other which are the most difficult to pin down. A great

[1] The great stumbling-block here for the internalist school of historiography of science (see below), is the question of historical causation. Scenting economic determinism under every formulation, they insist that the scientific revolution, as primarily a revolution in scientific ideas, cannot have been 'derivative from' some other social movement such as the Reformation or the rise of capitalism. Perhaps for the moment we could settle for some such phrase as 'indissolubly associated with. . . .' The internalists always seem to me essentially Manichaean; they do not like to admit that scientists have bodies, eat and drink, and live social lives among their fellow-men, whose practical problems cannot remain unknown to them; nor are the internalists willing to credit their scientific subjects with sub-conscious minds.

deal can be said about this, for example the vitally important role of the 'higher artisanate' and its acceptance into the company of educated scholars at this time;[1] but the present essay is not the place for it because we are in pursuit of something else. For us the essential point is that the development of modern science occurred in Europe and nowhere else.

In comparing the position of Europe with China the greatest and most obscure problems are (a) how far and in what way did Chinese medieval feudalism (if that is the proper term for it) differ from European feudalism, and (b) did China (or indeed India) ever pass through a 'slave society' analogous to that of classical Greece and Rome? The question is, of course, not merely whether the institution of slavery existed, that is quite a different matter, but whether the society was ever based on it.

In my early days, when I was still a working biochemist, I was greatly influenced by Karl A. Wittfogel's book *Wirtschaft und Gesellschaft Chinas*, written when he was a more or less orthodox Marxist in pre-Hitler Germany.[2] He was particularly interested in developing the conception of 'Asiatic bureaucratism', or 'bureaucratic feudalism', as I found later on that some Chinese historians called it. This arose from the works of Marx and Engels themselves, who had based it partly on, or derived it from, the observations of the seventeenth-century Frenchman François Bernier, physician to the Mogul emperor Aurangzeb in India.[3] Marx and Engels had spoken about the 'Asiatic mode of production'. How exactly they defined this at different times and how

[1] This factor was much emphasized and elaborated by the late Edgar Zilsel. Its importance has recently been recognized by a medievalist whom no one could suspect of Marxism, A. C. Crombie, in his 'The Relevance of the Middle Ages in the Scientific Movement' in *Perspectives in Mediaeval History*, ed. K. F. Drew & F. S. Lear, (Chicago, 1963) p. 35. See also his 'Quantification in Mediaeval Physics,' in *Quantification*, ed. H. Woolf, (Indianapolis, 1961) p. 13.

[2] Leipzig, 1931. I also learnt much from a golden little book by Hellmut Wilhelm, the son of the great sinologist Richard Wilhelm, *Gesellschaft und Staat in China* (Peiping, 1944). It is most unfortunate that this non-Marxist work has long been quite inaccessible, and that there has never been an English translation of it.

[3] *The History of the Late Revolution of the Empire of the Great Mogul*, originally published in French, Paris, 1671; many times republished, as by Dass, Calcutta, 1909. See the famous letter of Marx to Engels, 2 June 1853.

exactly it could or should be defined now is today once again the subject of animated discussions in nearly every country. Broadly speaking, it was the growth of a State apparatus fundamentally bureaucratic in character and operated by a non-hereditary élite upon the basis of a large number of relatively self-governing peasant communities, still retaining much tribal character and with little or no division of labour as between agriculture and industry. The form of exploitation here consisted essentially in the collection of taxes for the centralized State, i.e., the royal or imperial court and its regiments of bureaucratic officials. The justification

FIGURE 27   A Chinese picture of a flash-lock gate on a canal (bottom left). Although such water-controlling gates go back to the Han (first century) at least, pictures of them in traditional style are very rare. This one shows a lock-gate of stop-log type near Wu-Tha Ssu (the Five Pagoda Temple) just west of Peking, doubtless on the canal which fed the upper reaches of the most northerly section of the Grand Canal. The drawing comes from Wanyen Lin-Chhing's *Hung Hsüeh Yin Yuan Thu Chi* (Illustrated Account of the things that had to happen in my life), 1843, reproduced by Baylin.

of the State apparatus was, of course, twofold: on the one hand it organized the defence of the whole area (whether an ancient 'feudal' state or later the entire Chinese empire) and on the other hand it organized the construction and maintenance of public works. It is possible to say without fear of contradiction that throughout Chinese history the latter function was more important than the first, and this was one of the things that Wittfogel saw. The necessities of the country's topography and agriculture imposed from the beginning a vast series of water-works directed to (a) the conservation of the great rivers, in flood protection and the like, (b) the use of water for irrigation, especially for wet rice cultivation, and (c) the development of a far-flung canal system whereby the tax-grain could be brought to granary centres and to the capital. All this necessitated, besides tax exploitation, the organization of corvée labour, and one might say that the only duties of the self-governing peasant communities vis-à-vis the State apparatus were the payment of tax and the provision of labour power for public purposes when called upon to give it.[1] Besides this the State bureaucracy assumed the function of the general organization of production, i.e., the direction of broad agricultural policy, and for this reason the State apparatus of such a type of society is now receiving the appellation of 'an economic high command'. Only in China do we find among the most ancient high officials the Ssu Khung, the Ssu Thu and the Ssu Nung (Directors of the Multitudes for Engineering Works and for Agriculture). Nor can we forget that the 'nationalization' of salt and iron manufacture (the only commodities which had to travel, because not everywhere producible), suggested first in the fifth century B.C., was thoroughly put into practice in the second century B.C. Also in the Han period there was a governmental Fermented Beverages Authority; and there are many examples

[1] Today they do not have to give it, but are paid at the ordinary commune rate per labour day, and the work is done by the country people at slack times (A. L. Strong, *Letter from China*, 1964, No. 15). This principle of the rational and maximal utilization of man-power is one which goes back more than two thousand years in Chinese history, and its timing was one of the functions of the 'economic high command'.

of similar bureaucratic industries under subsequent dynasties.[1]

Various other aspects of this situation reveal themselves as one looks further into it; for example, peasant production was not under private control or ownership but public control, and theoretically all the land within the whole empire belonged to the Emperor and the Emperor alone. There was at first a semblance of landed property securely held by individual families, but this institution never developed in Chinese history in a way comparable with the feudal fief tenures of the West, since Chinese society did not retain the system of primogeniture. Hence all landed estates had to be parcelled out at each demise of the head of the family. Again, in that society, the conception of the city-state was totally absent; the towns were purposefully created as nodes in the administrative network, though very often no doubt they tended to grow up at spontaneous market centres. Every town was a fortified city held for the Prince or the Emperor by his civil governor and his military official. Since the economic function was so much more important in Chinese society than the military it is not surprising that the governor was usually a more highly respected personality than the garrison commander. Lastly, broadly speaking, slaves were not used in agricultural production, nor indeed very much in industry; slavery was primarily domestic, or as some would say, 'patriarchal' in character,[2] throughout the ages.

In its later highly developed forms such as one finds in Thang or Sung China the 'Asiatic mode of production' developed into a social system which while fundamentally 'feudal' in the limited sense that most of the wealth was based on agricultural exploitation,[3] was essentially bureaucratic and not military-aristocratic.

[1] Cf. H. F. Schurmann, *The Economic Structure of the Yuan Dynasty* (Cambridge, Mass., 1956), pp. 146 ff.

[2] See F. Tökei, 'Die Formen der chinesischen patriarchalischen Sklaverei in der Chou-Zeit,' in *Opuscula Ethnologica Memoriae Ludovici Biró Sacra*, (Budapest, 1959), p. 291.

[3] This must not be taken to mean that industry and trade were poorly developed in the Middle Ages. On the contrary, especially in the Southern Sung in the twelfth and thirteenth centuries, they were so productive and prosperous that the continuance of the typical bureaucratic forms is what surprises

It is quite impossible to overestimate the depth of the civilian *ethos* in Chinese history. Imperial power was exercised not through a hierarchy of enfeoffed barons but through an extremely elaborate civil service which Westerners know of as the 'mandarinate', enjoying no hereditary principle of succession to estates but recruited afresh in every generation. All I can say is that throughout nearly thirty years of study of Chinese culture these conceptions have made more sense in understanding Chinese society than any others. I believe that it will be possible to show in some considerable detail why the Asian 'bureaucratic feudalism' at first favoured the growth of natural knowledge and its application to technology for human benefit, while later on it inhibited the rise of modern capitalism and of modern science in contrast with the other form of feudalism in Europe which favoured it—by decaying and generating the new mercantile order of society. A predominantly mercantile order of society could never arise in Chinese civilization because the basic conception of the mandarinate was opposed not only to the principles of hereditary aristocratic feudalism but also to the value-systems of the wealthy merchants. Capital accumulation in Chinese society there could indeed be, but the application of it in permanently productive industrial enterprises was constantly inhibited by the scholar-bureaucrats, as indeed was any other social action which might threaten their supremacy. Thus, the merchant guilds in China never achieved anything approaching the status and power of the merchant guilds of the city-states of European civilization.

In many ways I should be prepared to say that the social and economic system of medieval China was much more rational than that of medieval Europe. The system of imperial examinations for entry into the bureaucracy, a system which had taken its origin as far back as the second century B.C., together with the age-old practice of the 'recommendation of outstanding talent', brought it about that the mandarinate recruited to itself the best brains of the nation (and the nation was a whole sub-continent) for more than two thousand years.[1] This stands in very great contrast

[1] A remarkable sidelight on this will be found in the paper by Lu Gwei-Djen & J. Needham, 'China and the Origin of (Qualifying) Examinations in Medicine', *Proc. Roy. Soc. Med.* 1963, **56**, 63.

FIGURE 28 A late Chhing illustration of the statement in the *Shu Ching* (Historical Classic), *c.* 9th century B.C.; 'If the ruler knows men, he is wise, and can nominate (the right) men for office,' (Kao Yao Mo ch., from *Shu Ching Thu Shuo*). The legendary emperor Shun is here seen with his eight benevolent ministers (Khai) and his eight honest vice ministers (Yuan).

to the European situation where the best brains were not especially likely to arise in the families of the feudal lords, still less among the more restricted group of eldest sons of feudal lords. There were of course certain bureaucratic features of early medieval European society, for example the office of the 'Counts', the institutions which gave rise to the position of 'Lord Lieutenant', and the widely customary use of bishops and clergy as administrators under the king, but all this fell far short of the systematic utilization of administrative talent which the Chinese system brought fully into play.

Moreover, not only was administrative talent brought forward and settled thoroughly into the right place but so strong was the Confucian *ethos* and ideal that the chief representatives of those who were not scholar-gentry remained for the most part conscious of their lesser position in the scheme of things. When I was giving a talk recently to a university society on these subjects someone asked the excellent question, 'How was it that the military men could accept their inferiority to the civil officials throughout Chinese history?' After all, 'the power of the sword' has been overwhelming in other civilizations. What immediately came to my mind in replying was the imperial *charisma* carried by the bureaucracy,[1] the holiness of the written character (when I first went to China the stoves for giving honourable cremation to any piece of paper with words written on it were still to be seen in every temple), and the Chinese conviction that the sword might win but only the *logos* could keep. There is a famous story about the first Han emperor who was impatient with the elaborate ceremonies devised for the court by his attendant philosophers, till one of them said to him, 'You conquered the empire on horseback, but from horseback you will never succeed in ruling it.' Thereafter the rites and ceremonies were allowed to unfold in all their liturgical majesty.[2] In ancient times the Chinese leader was often an important official and a general indiscriminately

[1] One should add the high moral standards of Confucianism which exerted great social pressure throughout the ages upon the members of the mandarinate.
[2] See *SCC*, Vol. I, p. 103.

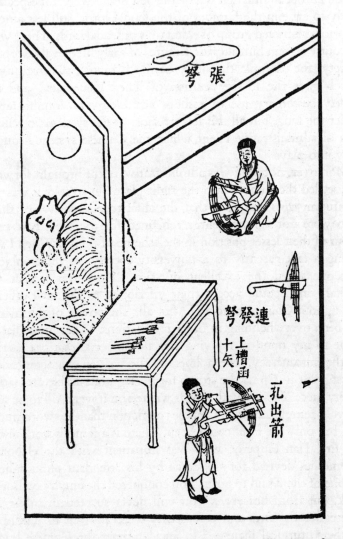

弩張

連發弩
上槽函
十矢
孔出箭

FIGURE 29 Makers of crossbows in their workshop, an illustration from the first edition of the *Thien Kung Khai Wu*, 1637. In the foreground a 'machine-gun' crossbow is being tested; this was a development which did not occur in Europe. The captions say that the magazine holds ten bolts or quarrells which are shot off through the orifice in front, each one being brought into position automatically by the act of arming.

and what is significant is that the psychology of military men themselves clearly admitted their inferiority. They were very often 'failed civilians'. Of course, force was the ultimate argument, the final sanction, as in all societies, but the question was—what force, moral or purely physical? The Chinese profoundly believed that only the former lasted, and what the latter could gain only the former could keep.

Furthermore, there may have been technical factors in the primacy of the spoken and written word in Chinese society. It has been demonstrated that in ancient times the progress of invention in offensive weapons, especially the efficient crossbow, far outstripped progress in defensive armour. There are many cases in antiquity of feudal lords being killed by commoners or peasants well armed with crossbows—a situation quite unlike the favourable position of the heavily armed knight in Western medieval society. Hence, perhaps, arose the Confucian emphasis on persuasion. The Chinese were Whigs, 'for Whigs do use no force but argument'. The Chinese peasant-farmer could not be driven into battle to defend the boundaries of his State, for instance, before the unification of the Empire, since he would be quite capable of shooting his Prince first; but if he was persuaded by the philosophers, whether patriots or sophists, that it was necessary to fight for that State, as, indeed, also later for the Empire, then he would march. Hence, the presence of a certain amount of what one might call 'propaganda' (not necessarily in a pejorative sense) in Chinese classical and historical texts—a kind of 'personal equation' for which the historian has to make proper allowance. There was nothing peculiar to China in this. It is, of course, a world-wide phenomenon notable from Josephus to Gibbon, but the sinologist has always to be on the lookout for it, for it was the *défaut* of the civilized civilian *qualité*.

Yet another argument is of interest in this connection, namely the fact that the Chinese was always primarily a peasant-farmer, and not engaged in either animal husbandry or sea-faring.[1] These two latter occupations encourage excessive command and obedi-

[1] This contrast was, I think, first appreciated by André Haudricourt.

ence; the cowboy or shepherd drives his animals about, the sea-captain gives orders to his crew which are neglected at the peril of everyone's life on board, but the peasant-farmer, once he has done all that is necessary for the crops, must wait for them to come up. A famous parable in Chinese philosophical literature derides a man of Sung State who was discontented with the growth rate of his plants and started to pull at them to help them to come up.[1] Force, therefore, was always the wrong way of doing things, hence civil persuasion rather than military might was always the correct way of doing things. And everything that one could say for the position of the soldier vis-à-vis the civil official holds good, *mutatis mutandis*, for the merchant. Wealth as such was not valued. It had no spiritual power. It could give comfort but not wisdom, and in China affluence carried comparatively little prestige. The one idea of every merchant's son was to become a scholar, to enter the imperial examination and to rise high in the bureaucracy. Thus did the system perpetuate itself through ten thousand generations. I am not sure that it is still not alive, though raised of course to a higher plane, for does not the Party official, whose position is quite irrelevant to the accidents of his birth, despise both aristocratic values on the one hand and acquisitive values on the other? In a word, perhaps socialism was the spirit of un-dominating justice imprisoned within the shell of Chinese medieval bureaucratism.[2] Basic Chinese traditions may perhaps be more congruent with the scientific world co-operative commonwealth than those of Europe.

Between 1920 and 1932 there were great discussions in the Soviet Union about what Marx had meant by the 'Asiatic mode of production', but very little is known of these in Western countries because they were never translated. If any copies of the Russian accounts still exist it would be highly desirable to have them re-

[1] See *SCC*, Vol. II, p. 576.
[2] Of course the medieval mandarinate was part of an exploiting system, like those of Western feudalism and capitalism, but as a non-hereditary élite it did oppose both aristocratic and mercantile ways of life. Cf. the work by C. Brandt, B. Schwartz & J. K. Fairbank, *A Documentary History of Chinese Communism* (Cambridge, Mass., 1952), and J. Needham, 'The Past in China's Present,' in *Within the Four Seas* (London, 1969).

published in Western languages. Although we have never been able to study the results, it is believed that those who opposed any variation from the standard succession, primitive communalism—slave-society—feudalism—capitalism—socialism, gained the day. No doubt the climate of dogmatism which prevailed in the social sciences during the personality-cult period played some role in this situation.[1] We now have younger writers expressing the great embarrassment felt by English Marxists that 'feudalism' has become a meaningless term.[2] 'Obviously,' they say, 'a socio-economic stage which covers both Ruanda-Urundi today and France in 1788, both China in 1900 and Norman England, is in danger of losing any kind of specific character likely to assist analysis. . . .' Sub-divisions are desperately needed. The remarkable thing is that these writers do not seem to know much about the original views of Marx and Engels. 'The "Asiatic mode," ' one of them says, 'has long since tacitly dropped out of use.'[3] However, the same writer goes on to pose the problem of the arrested development of certain Asian and African societies very well, and recommends the 'rehabilitation of Marx's "Asiatic mode" or even several modes, to enable' a differentiation in nomenclature between regional variations. The use of the term 'proto-feudal' (which I believe I invented myself) is also recommended for a single basic stage which then developed in different ways.

Whenever the name of Wittfogel appears in Marxist writings nowadays it is mentioned with aversion. The reason for this is that during the Hitler period Wittfogel migrated to America, where he still works. For many years he has been a great brandisher of tomahawks in the intellectual Cold War, and those writers who regard his recent book *Oriental Despotism*[4] as propaganda directed

---

[1] During the subsequent decades there have been many distinguished sociological studies of Asian cultures by Russian sinologists, usually avoiding, however, the concept of the 'Asiatic mode of production.'

[2] J. Simon, in *Marxism Today*, 1962, 6 (no. 6), 183.   [3] J. Simon, loc. cit.

[4] New Haven, 1957. Reviewed, among other things, by J. Needham, *Science & Society*, 1959, 23, 58. Among the many critiques of Wittfogel's ideas may be mentioned an interesting recent study from the juristic point of view by Orlan Lee, 'Traditionelle Rechtsgebräuche und der Begriff d. Orientalischer Despotismus,' *Zeitschr. J. vergl. Rechtswiss.*, 1964, 66, 157.

against Russia and China both old and new are only too probably correct. Wittfogel now seeks to attribute all abuses of power, whether in totalitarian or other societies, to the principle of bureaucratism; but the fact that he has become a great opponent of the ideas which I and many others favour does not alter the fact that he once set them forth quite brilliantly himself, and thus I admire his first book while deprecating his last one. Although Wittfogel has perhaps overdone it, I do not regard his theory of 'hydraulic society' as essentially erroneous, for I also believe that the spatial range of public works (river control, irrigation and the building of transport canals) in Chinese history transcended time after time the barriers between the territories of individual feudal or proto-feudal lords. It thus invariably tended to concentrate power at the centre, i.e., in the bureaucratic apparatus arched above the granular mass of 'tribal' clan villages. I think it played an important part, therefore, in making Chinese feudalism 'bureaucratic'. Of course it does not matter from the standpoint of the historian of science and technology how different Chinese feudalism was from European feudalism, but it has got to be different enough (and I firmly believe that it was different enough) to account for the total inhibition of capitalism and modern science in China as against the successful development of both these features in the West.

As for bureaucracy, it is sheer nonsense to lay all social evil at its door. On the contrary it has been through the ages a magnificent instrument of human social organization. Furthermore, it is going to be with us, if humanity endures, for many centuries to come. The fundamental problem before us now is the humanization of bureaucracy, so that under socialism not only shall its organizing power be used for the benefit of the ordinary man and woman; but that it shall be known and palpably felt and seen to be so used. Modern human society is, and will increasingly be, based on modern science and technology, and the more this goes on the more indispensable a highly organized bureaucracy will be. The fallacy here is to compare such a system after the rise of modern science with *any* precursor systems which existed before it. For modern science has given us a vast wealth of instruments from telephones to computers which now and only now could truly implement the will to humanize bureaucracy. That will may rest on what is essentially Confucianism, Taoism, and revolutionary Christianity as well as Marxism.

The term 'oriental despotism' recalls of course the speculations of the Physiocrats in eighteenth-century France, who were deeply influenced by what was then known of the Chinese economic and social structure.[1] For them, of course, it was an enlightened despotism, which they much admired, not the grim and wicked system of Wittfogel's later imagination. Sinologists throughout the world were impatient with his later book[2] because it persistently selected from the facts. Thus, for example, it is impossible to say that there was no educated public opinion in medieval China. On the contrary the scholar-gentry and the scholar-bureaucrats constituted a wide and very powerful public opinion, and there were times when the Emperor might command but the bureaucracy would not obey.[3] In theory, the Emperor might be an absolute ruler, but in practice what happened was regulated by long-established precedent and convention, interpreted age after age by the Confucian exegesis of historical texts. China has always been a 'one-Party State', and for over two thousand years the rule was that of the Confucian Party. My opinion is, therefore, that the term 'oriental despotism' is no more justified in the hands of Wittfogel than it was in those of the Physiocrats, and I never use it myself. On the other hand there are many Marxist terms, some old and some now gaining prominence, which I find great difficulty in adopting. For example, in some texts the 'imaginary State construct' is contrasted with the 'real substratum' of the independent peasant villages. This does not seem to me justifiable because in its way the State apparatus was quite as real as the work

[1] On this see L. A. Maverick's *China a Model for Europe* (San Antonio, Texas, 1946) which includes a translation of F. Quesnay's *Le Despotisme de la Chine*, (Paris, 1767).

[2] See, for example, the review by E. G. Pulleyblank in *Bulletin of the London School of Oriental and African Studies*, 1958, **21**, 657.

[3] Cf. Liu Tzu-Chien, 'An Early Sung Reformer, Fan Chung-Yen,' in *Chinese Thought and Institutions*, ed. J. K. Fairbank (Chicago, 1957), p. 105.

of the peasant-farmers. Nor do I like to apply the term 'autonomic' to the village communities because I think it was only true within very definite limitations. The truth is that we urgently need the development of some entirely new technical terms. We are dealing here with states of society far removed from anything that the West ever knew, and in coining these new technical terms I would suggest that we might make use of Chinese forms rather than continuing to insist on using Greek and Latin roots to apply to societies which were enormously different. Here the term *kuan-liao* for the bureaucracy might come in useful. If we could get a more adequate terminology it would also help us to consider certain other related problems. Here I am thinking of the remarkable fact that Japanese society was more similar to that of Western European society, and for that very reason more capable of developing modern capitalism. This has been recognized by historians for a long time past, but recent writings have pinpointed rather precisely the exact ways in which Japanese military-aristocratic feudalism could generate capitalism as Chinese bureaucratic society could not.[1]

Next, I may say something, though it will not be much, about 'slave-society'. According to my own experiences with Chinese archaeology and literature, for what they are worth, I am not very inclined to believe that Chinese society, even during the Shang and early Chou periods, was ever a slave-based society in the same sense as the Mediterranean cultures with their slave-manned galleys plying the Mediterranean and their *latifundia* spread over the fields of Italy. Here I diverge, with deep humility, from some contemporary Chinese scholars, who have been extremely impressed by the 'single-track' system of developmental stages of society prominent in Marxist thinking during the past twenty or thirty years. The subject is still under intensive debate and we cannot yet say that certainty has been achieved in any aspect of it.

[1] See, for example, the recent monograph by N. Jacobs, *The Origin of Modern Capitalism and Eastern Asia* (Hongkong, 1958), notable also for the excellence of its index. The author is a Weberian sociologist who executes the remarkable feat of making no mention of Marx and Engels. Evidently the Departments of History of Economics and History of Science at Hongkong inhabit separate ivory pagodas.

Some years ago at Cambridge we had a symposium on slavery in the different civilizations, in the course of which the participants all had to agree that the actual forms of slavery were very different in Chinese society from anything known elsewhere. Owing to the dominance of clan and family obligations it was rather doubtful whether anyone in that civilization could have been called 'free' in some of the Western senses, while on the other hand (contrary to what many believe) chattel-slavery was distinctly rare.[1] The fact is that no one really yet knows fully what was the status of servile and semi-servile groups in the different periods in China (and there were many very different kinds of such groups), neither Western sinologists nor even the Chinese scholars themselves. A great amount of research remains to be done, but I think it seems already clear that neither in the economic nor in the political field was chattel-slavery ever a basis for the whole of society in China in the same way as it was at some times in the West.

Although the question of the slave basis of society has a certain importance in so far as it affects the position of science and technology among the Greeks and Romans, it is of course less germane to what was originally my central point of interest, namely the origin and development of modern science in the late Renaissance in the West. It could, however, have a very important bearing on the greater success of Chinese society in the application of the sciences of Nature to human benefit during the earlier period, the first fourteen centuries of the Christian era and four or five centuries prior to that. Is it not very striking and significant that China has nothing whatever to show comparable with the use of slaves at sea in galleys in the Mediterranean? Sail, and a very refined use of it, was the universal method of propulsion of Chinese ships from ancient times. China has no records of the mass use of the human motor comparable with the building methods of ancient Egypt. So also it is remarkable that we have never so far come across any important instance of the refusal of an invention in Chinese society due to fear of technological unemployment. If

[1] See E. G. Pulleyblank, 'The Origins and Nature of Chattel-Slavery in China', *Journal of Economic and Social History of the Orient*, 1958, **1**, 185.

Chinese labour-power was so vast as most people imagine, it is not easy to see why this factor should not sometimes have come into play. We have numerous examples of labour-saving devices introduced at early times in Chinese culture, very often much earlier than in Europe. A concrete case would be the wheelbarrow, not known in the West before the thirteenth century but common in China in the third and arising there almost certainly two hundred years earlier than that. It may well be that just as the bureaucratic apparatus will explain the failure of modern science to arise spontaneously in Chinese culture, so also the absence of mass chattel-slavery may turn out to have been an important factor in the greater success of Chinese culture in fostering pure and applied science in the earlier centuries.

At the present time there is a real ferment among younger European sociological scholars concerning the reconsideration of the problem of the 'Asiatic mode of production.'[1] This may have been brought about partly because of the importance of such ideas for the interpretation of African societies now emerging from underdeveloped conditions. It is not clear that the restricted categories which had become conventional will altogether account for them, but the greatest stimulus perhaps has been the publication in Moscow in 1939 of a text of Marx himself written in 1857 and 1858 entitled *Formen die der kapitalistischen Produktion vorhergehen*. This was a kind of trial run for his *Kapital*, and is included in his *Grundrisse der Kritik der politischen Ökonomie*, a collection of basic papers published again in Germany in 1952.[2] It was singularly unfortunate that Marx's text was not known to the participants in the Russian discussions of the twenties and thirties, for it gave the only deep and systematic exposé of his ideas on the 'Asiatic mode of production'. One great question is whether Marx and Engels regarded this as something qualitatively different from one or other of the classically distinguished types of society in the rest of

---

[1] See especially the review by J. Chesneaux, 'La Mode de Production Asiatique; une nouvelle Étape de la Discussion,' *Eirene*, 1964, as well as the several valuable contributions in *Pensée* 1964 (no. 114).

[2] Dietz, Berlin. An English edition has now appeared.

the world, or only quantitatively different. It is not yet clear whether they saw it as essentially a 'transitory' situation (though in some cases it might be capable of age-long stabilization) or whether they thought of 'bureaucratism' as a fourth fundamental type of society. Was the 'Asiatic mode of production' simply a variation of classical feudalism? Some Chinese historians have certainly regarded it as a special type of feudalism. But sometimes Marx and Engels seemed to speak as if they did consider it as something qualitatively different from slave production or feudal production. There was also always the question how far the conceptions of 'bureaucratic feudalism' might be applicable to pre-Columbian America or other societies such as medieval Ceylon. This is the kind of problem to which Wittfogel has applied himself much in recent times but without satisfactory conclusions (Ceylon is not even mentioned in his index), and the younger sociologists are now attacking it in quite a different light.[1]

I have no doubt that their thinking will greatly elucidate my own problem of the early advanced and later retarded character of Chinese science and technology. In particular my French friends and colleagues Jean Chesneaux and André Haudricourt have been active in the matter, and what follows is based on some ideas which they have presented to me. It seems clear that the early superiority of Chinese science and technology through long centuries must be placed in relation with the elaborate, rationalized and conscious mechanisms of a society having the characters of 'Asiatic bureaucracy'. It was a society which functioned fundamentally in a 'learned' way, the seats of power being filled by scholars, not military commanders. Central authority relied a great deal upon the 'automatic' functioning of the village communities, and in general tended to reduce to the minimum its intervention in their life. I have already spoken of the fundamental difference between peasant-farmers on the one hand and shepherds or seamen on the other. This difference is expressed epigrammatically in the Chinese

[1] On the Ceylonese situation, where hydraulic works were very remarkable and abundant but generated no mandarinate, see E. R. Leach, 'Hydraulic Society in Ceylon', *Past & Present*, 1959 (no. 15).

terms *wei* and *wu wei*. *Wei* meant the application of force, of will-power, the determination that things, animals or even other men, should do what they were ordered to do; but *wu wei* was the opposite of this, leaving things alone, letting Nature take her course, profiting by going with the grain of things instead of going against it, and knowing how not to interfere. This was the great Taoist watchword throughout the ages, the untaught doctrine, the wordless edict.[1] It was summarized in that numinous phrase which Bertrand Russell collected from his time in China, 'production without possession, action without self-assertion, development without domination'.[2] Now *wu wei*, the lack of interference, might very well be applied to a respect for the 'automotive' capacity of the individual farmers and their peasant communities. Even when the old 'Asiatic' society had given place to 'bureaucratic feudalism' such conceptions remained very much alive. Chinese political practice and government administration was continually founded upon that non-intervention inherited from ancient Asian society and from the single pair of opposites 'villages-prince'. Thus, all through Chinese history, the best magistrate was he who intervened least in society's affairs, and all through history, too, the chief aim of clans and families was to settle their affairs internally without having recourse to the courts.[3] It seems probable that a society like this would be favourable to reflection upon the world of Nature. Man should try to penetrate as far as possible into the mechanisms of the natural world and to utilize the sources of power which it contained while intervening directly as little as possible, and utilizing 'action at a distance'. Conceptions of this kind, highly intelligent, sought always to achieve effects with an economy of means, and naturally encouraged the investigation of Nature for essentially Baconian reasons. Hence such early triumphs as those of the seismograph, the casting of iron, and water-power.

[1] Cf. *SCC*, Vol. II, p. 564.
[2] *SCC*, Vol. II, p. 164; from *The Problem of China* (London, 1922), p. 194.
[3] An aspect of the darker side of this is given in the partly autobiographical account of my old friend Kuo Yu-Shou, *La Lune sur le Fleuve Perle*, (Paris, 1963).

It might thus be said that this non-interventionist conception of human activity was, to begin with, propitious for the development of the natural sciences. For example, the predilection for 'action at a distance' had great effects in early wave-theory, the discovery of the nature of the tides, the knowledge of relations between mineral bodies and plants as in geo-botanical prospecting, or again in the science of magnetism. It is often forgotten that one of the fundamental features of the great break-through of modern science in the time of Galileo was the knowledge of magnetic polarity, declination, etc.; and unlike Euclidean geometry and Ptolemaic astronomy magnetical science had been a totally non-European contribution.[1] Nothing had been known of it to speak of in Europe before the end of the twelfth century, and its transmission from the earlier work of the Chinese is not in doubt. If the Chinese were (apart from the Babylonians) the greatest observers among all ancient peoples, was it not perhaps precisely because of the encouragement of non-interventionist principles, enshrined in the numinous poetry of the Taoists on the 'water symbol' and the 'eternal feminine'?[2]

However, if the non-interventionist character of the 'villages-prince' relationship engendered a certain conception of the world which was propitious to the progress of science, it had certain natural limitations. It was not congruent with characteristically occidental 'interventionism', so natural to a people of shepherds and sea-farers. Since it was not capable of allowing the mercantile mentality a leading place in the civilization, it was not capable of fusing together the techniques of the higher artisanate with the methods of mathematical and logical reasoning which the scholars had worked out, so that the passage from the Vincian to the Galilean stage in the development of modern natural science was not achieved, perhaps not possible. In medieval China there had been more systematic experimentation than the Greeks had ever

[1] See J. Needham, 'The Chinese Contribution to the Development of the Mariner's Compass,' *Scientia*, 1961, **55**, 1; *Actas de Congresso Internacional de História des Descobrimentos* (Lisbon, 1961), Vol. II, p. 311.

[2] Cf. *SCC*, Vol. II, p. 57.

attempted, or medieval Europe either, but so long as 'bureaucratic feudalism' remained unchanged, mathematics could not come together with empirical Nature-observation and experiment to produce something fundamentally new. The suggestion is that experiment demanded too much active intervention, and while this had always been accepted in the arts and trades, indeed more so than in Europe, it was perhaps more difficult in China to make it philosophically respectable.

There was another way, also, in which medieval Chinese society had been highly favourable to the growth of the natural sciences at the pre-Renaissance level. Traditional Chinese society was highly organic, highly cohesive. The State was responsible for the good functioning of the entire society, even if this responsibility was carried out with the minimum intervention. One remembers that the ancient definition of the Ideal Ruler was that he should sit simply facing the south and exert his virtue (*tê*) in all directions so that the Ten Thousand Things would automatically be well-governed. As we have been able to show over and over again, the State brought powerful aid to scientific research.[1] Astronomical observatories, for example, keeping millennial records, were part of the civil service, vast encyclopaedias not only literary but also medical and agricultural were published at the expense of the State, and scientific expeditions altogether remarkable for their time were successfully accomplished (one thinks of the early eighth-century geodetic survey of a meridian arc stretching from Indo-China to Mongolia, and the expedition to chart the constellations of the southern hemisphere to within twenty degrees of the south celestial pole).[2] By contrast science in Europe was generally a private enterprise. Therefore it hung back for many centuries. Yet, the State science and medicine of China was not capable of making, when the time came, that qualitative leap which happened in occidental science in the sixteenth and early seventeenth centuries.

[1] *SCC*, Vols. II, III, IV, VI *passim*.
[2] See A. Beer, Ho Ping-Yü, Lu Gwei-Djen, J. Needham, E. G. Pulleyblank & G. I. Thompson, 'An Eighth-Century Meridian Line; I-Hsing's Chain of Gnomons and the Prehistory of the Metric System,' *Vistas in Astronomy*, 1961, **4**, 3.

Some Asian scholars have been suspicious of the idea of the 'Asiatic mode of production' or 'bureaucratic feudalism' because they have identified it with a certain 'stagnation' which they thought they saw in the history of their own societies. In the name of the right of the Asian and African peoples to progress they have projected this feeling into the past and have wished to claim for their ancestors exactly the same stages as those which the West had itself gone through, that Western world which had for a time dominated so hatefully over them. It is, I think, very important to clear up this misunderstanding, for there seems no reason at all why we should assume *a priori* that China and other ancient civilizations passed through exactly the same social stages as the European West. In fact, the word 'stagnation' was never applicable to China at all; it was purely a Western misconception. A continuing general and scientific progress manifested itself in traditional Chinese society, but this was violently overtaken by the exponential growth of modern science after the Renaissance in Europe. China was homoeostatic, cybernetic if you like, but never stagnant. In case after case it can be shown with overwhelming probability that fundamental discoveries and inventions made in China were transmitted to Europe, for example, magnetic science, equatorial celestial co-ordinates and the equatorial mounting of observational astronomical instruments,[1] quantitative cartography, the technology of cast iron,[2] essential components of the reciprocating steam-engine such as the double-acting principle and the standard interconversion of rotary and longitudinal motion,[3] the mechanical clock,[4] the boot stirrup and the efficient equine harnesses, to say nothing of gunpowder and all that followed there-

[1] J. Needham, 'The Peking Observatory in 1280 and the Development of the Equatorial Mounting,' *Vistas in Astronomy*, 1955, **1**, 67.

[2] Cf. J. Needham, *The Development of Iron and Steel Technology in China* (London, 1958).

[3] See my Earl Grey lecture at the Newcastle University, 1961, "Classical Chinese Contributions to Mechanical Engineering,' and my Newcomen Centenary Lecture, 'The Pre-Natal History of the Steam-Engine,' *Trans. Newcomen Soc.*, 1962, **35**, 3.

[4] Cf. J. Needham, Wang Ling & D. J. de S. Price, *Heavenly Clockwork* (Cambridge, 1960).

from.[1] These many diverse discoveries and inventions had earth-shaking effects in Europe, but in China the social order of bureaucratic feudalism was very little disturbed by them. The built-in instability of European society must therefore be contrasted with a homoeostatic equilibrium in China, the product I believe of a society fundamentally more rational. What remains is an analysis of the relationships of social classes in China and Europe. The clashes between them in the West have been charted well enough, but in China the problem is much more difficult because of the non-hereditary nature of the bureaucracy. This is a task for the future.

In recent decades much interest has been aroused in the history of science and technology in the great non-European civilizations, especially China and India, interest, that is, on the part of scientists, engineers, philosophers and orientalists, but not, on the whole, among historians. Why, one may ask, has the history of Chinese and Indian science been unpopular among them? Lack of the necessary linguistic and cultural tools for approaching the original sources has naturally been an inhibition, and of course if one is primarily attracted by eighteenth- and nineteenth-century science European developments will monopolize one's interest. But I believe there is a deeper reason.

The study of great civilizations in which *modern* science and technology did not spontaneously develop obviously tends to raise the causal problem of how modern science did come into being at the European end of the Old World, and it does so in acute form. Indeed, the more brilliant the achievements of the ancient and medieval Asian civilizations turn out to have been the more discomforting the problem becomes. During the past thirty years historians of science in Western countries have tended to reject the sociological theories of the origin of modern science which had a considerable innings earlier in this century. The forms in which such hypotheses had then been presented were doubtless

[1] Some of the multifarious influences of Chinese inventions and discoveries on the pre-Renaissance world have been emphasized by Lynn White in his *Medieval Technology and Social Change* (Oxford, 1962).

relatively crude,[1] but that was surely no reason why they should not have been refined. Perhaps also the hypotheses themselves were felt to be too unsettling for a period during which the history of science was establishing itself as a factual academic discipline. Most historians have been prepared to see science having an influence on society, but not to admit that society influenced science, and they have liked to think of the progress of science solely in terms of the internal or autonomous filiation of ideas, theories, mental or mathematical techniques and practical discoveries, handed on like torches from one great man to another. They have been essentially 'internalists' or 'autonomists'. In other words, 'there was a man sent from God, whose name was. . . .' Kepler.[2]

The study of other civilizations, therefore, places traditional historical thought in a serious intellectual difficulty. For the most obvious and necessary kind of explanation which it demands is one which would demonstrate the fundamental differences in social and economic structure and mutability between Europe on the one hand and the great Asian civilizations on the other, differences which would account not only for the development of modern

[1] Such is the adjective generally applied to B. Hessen's famous paper 'On the Social and Economic Roots of Newton's *Principia*,' delivered at the International Congress of the History of Science at London in 1931 (reprinted in *Science at the Cross-roads*, London, 1932). It was certainly in plain blunt Cromwellian style. But already half a dozen years later, R. K. Merton's remarkable monograph, 'Science, Technology and Society in Seventeenth-Century England,' *Osiris*, 1938, **4**, 360–632, had achieved a considerably more refined and sophisticated presentation. Much is owing also to the works of E. Zilsel, several of which were published in the *Journal of the History of Ideas*, and all of which ought to be collected in a single volume.

[2] Though off the rails at various points, J. Agassi is entertaining on this topic in his monograph 'Towards a Historiography of Science' in *History and Theory* (1963), Beiheft 2. The 'inductivist' historians of science, he says, are chiefly concerned with questions of whom to worship and for what reasons; but he does not like the 'conventionalists' much better. With this particular quarrel I am not here concerned, but it is surprising that Agassi did not make more use of the work of Walter Pagel, which would have supported some of his arguments strongly. On the whole, Agassi takes his own stand for autonomism, regarding Marxism as one of the failings of inductivists, and believing that contention between different schools was the main factor in the development of science. As his monograph comes from the University of Hongkong he seems to have encysted himself with extraordinary success from all contact with Chinese culture—at any rate, so far.

science in Europe alone, but also of capitalism in Europe alone, together with its typical accompaniments of protestantism, nationalism, etc. not paralleled in any other part of the globe. Such explanations are, I believe, capable of much refinement. They must in no way neglect the importance of a multitude of factors in the realm of ideas—language and logic, religion and philosophy, theology, music, humanitarianism, attitudes to time and change—but they will be most deeply concerned with the analysis of the society in question, its patterns, its urges, its needs, its transformations. On the internalist or autonomist view such explanations are unwelcome. Those who hold it, therefore, instinctively dislike the study of the other great civilizations.

But if you reject the validity or even the relevance of sociological accounts of the 'scientific revolution' of the late Renaissance, which brought modern science into being, if you renounce them as too revolutionary for that revolution, and if at the same time you wish to explain why Europeans were able to do what Chinese and Indians were not, then you are driven back upon an inescapable dilemma. One of its horns is called pure chance, the other is racialism however disguised. To attribute the origin of modern science entirely to chance is to declare the bankruptcy of history as a form of enlightenment of the human mind. To dwell upon geography and harp upon climate as chance factors will not save the situation, for it brings you straight into the question of city-States, maritime commerce, agriculture and the like—concrete factors with which autonomism declines to have anything to do. The 'Greek miracle', like the scientific revolution itself, is then doomed to remain miraculous. But what is the alternative to chance? Only the doctrine that one particular group of peoples, in this case the European 'race', possessed some intrinsic superiority to all other groups of peoples. Against the scientific study of human races, physical anthropology, comparative haematology and the like, there can, of course, be no objection, but the doctrine of European superiority is racialism in the political sense and has nothing in common with science. For the European autonomist, I fear, 'we are the people, and wisdom was born with us'. However, since racialism (at least in its explicit forms) is neither intellectually respectable nor internationally acceptable, the autonomists are in a quandary which may be expected to become more obvious as time goes on.[1] I confidently anticipate therefore a great revival of interest in the relations of science and society during the crucial European centuries, as well as a study ever more intense of the social structures of all the civilizations, and the delineation of how they differed in glory, one from another.

In sum, I believe that the analysable differences in social and economic pattern between China and Western Europe will in the end illuminate, as far as anything can ever throw light on it, both the earlier predominance of Chinese science and technology and also the later rise of modern science in Europe alone.

---

[1] D. J. de S. Price, a valued collaborator of our own, knows much of the Asian contribution, but in his *Science Since Babylon* (New Haven, Conn., 1961) follows a 'hunch' of Einstein's and favours chance combinations of circumstances as the evocators of Greek and Renaissance science. A. R. Hall, in 'Merton Revisited', *History of Science*, 1963, **2**, 1, attacks anew what he calls the 'externalist' historiography of science, but significantly keeps silence about the problem posed by the Asian contributions. If he had taken a broader comparative point of view, his arguments about the European situation might have carried more conviction. A. C. Crombie (see p. 193), alone of the three, shows a real consciousness of the slow social changes which permitted the intellectual movements of the late Middle Ages and the Renaissance to bring modern science into being in the European culture-area, but even he pays less attention to their economic concomitants.

7

TIME AND EASTERN MAN

Henry Myers Lecture, Royal Anthropological Institute, 1964, also in
The Voices of Time, ed. J. T. Fraser (New York, 1966).

In this lecture my aim will be to describe the prevailing attitudes
to time in Chinese civilization.[1] I want to examine the claims which
have been made for Europe as the only culture with a real sense of
history, and to enquire whether, if that should truly be the case,
it could have had anything to do with the rise of modern science
and technology at the Renaissance and the time of the 'scientific
revolution'.

The *philosophia perennis* of Chinese culture was an organic naturalism which invariably accepted the reality and importance of time.
This must be related to the fact that although metaphysical ideal-
ism is found in China's philosophical history, and even enjoyed
occasionally a certain success, as when Buddhism was dominant
in the Liu Chhao and Thang periods, or among the followers of

[1] The title of this excursion into comparative anthropology is taken on the
rebound from that of a book which made a great literary stir in the twenties, the
'Time and Western Man' of P. Wyndham Lewis. This was a polemic against what
the writer considered the undue preoccupation with the flow of time in much
twentieth-century literature (Proust, Joyce, etc.), but it also contained a philosophical section in which he took issue with Einstein, Spengler, A. N. Whitehead and
Samuel Alexander, his particular *bête noire* being Bergson. This part, though the
larger, would hardly be taken seriously by philosophers today, if indeed it ever was,
but I am told that Wyndham Lewis's literary criticism is still highly regarded.
In any case I make him a bow for the benefit of title. I think I shall be able to show
that Western Man had no monopoly of the sense of linear continuous time, and
that the 'timeless Orient' is nonsense.

218

Wang Yang-Ming (A.D. 1472 to 1529) it never really occupied more than a subsidiary place in Chinese thinking. Subjective conceptions of time were therefore uncharacteristic of Chinese thought. Although we are speaking here, of course, of ancient and medieval or traditional thought, and not of sophisticated modern ideas, it may also be said that clear adumbrations of relativism occur in the ancient Taoist thinkers. But whatever happened in time, or times, whether flourishing or decay, time itself remained inescapably real for the Chinese mind. This contrasts strongly with the general ethos of Indian civilization,[1] and aligns China rather with the inhabitants of that other area of temperate climate at the western end of the Old World.

TIME IN CHINESE PHILOSOPHY AND
NATURAL PHILOSOPHY

Time and its content were often the subjects of discussion and speculation in the philosophical schools of the Warring States period (c. fourth century B.C.), contemporary with Aristotle.[2] We may take a brief look at what interested them. The expression which is now used for 'the universe', yü-chou, has essentially the meaning of 'space-time'. In a text of 120 B.C. we read:[3]

All the time that has passed from antiquity until now is called chou; all the space in every direction, above and below, is called yü. The Tao (the Order of Nature) is within them, yet no man can say where it dwells.'

The original meaning of both these two ancient words was

---

[1] Cf. H. Zimmer, Philosophies of India (New York, 1953), p. 450.

[2] No systematic treatment of this specific subject is available, so far as I know, either in Chinese or a Western language; a good deal of research would be necessary but the result would be well worth while. Marcel Granet only touched upon the question in his La Pensée Chinoise (Paris, 1934), pp. 90 ff., drawing mostly from non-philosophical sources, and in his time the analysis of ancient Chinese philosophy was much less advanced than it is now. The subject is not an easy one, for apart from t intrinsic difficulty of translating the arguments and opinions of the writers, their te... have often come down to us in a relatively corrupt state, so that even the quotations given here cannot as yet be regarded as more than provisional interpretations.

[3] Huai Nan Tzu, ch. 11.

'roof', of house, cart or boat, so that the semantic significance is that of something stretching over an expanse to cover it. So indeed we still in English say that such and such an exposition 'covers' ten or fifteen centuries. The word for duration (*chiu*) is explained by the Han lexicographers as derived from the character *jen*, man, a man stretching his legs and walking 'a stretch', just as a roof stretches across a space, and time stretches from one event to another.

Interesting definitions of time and space are contained in the writings which have come down to us from the Mohist school (Mo Chia), the followers of Mo Ti (*fl.* between 479 and 381 B.C.). This was the group of ancient Chinese thinkers most interested in the philosophy of mathematics and science. The *Mo Tzu* book is of different dates, for the systematic account of Master Mo's doctrine, including that of universal love (*chien ai*), cannot be much later than 400 B.C., while the Canons and their Expositions (a kind of corpus of definitions explained by a commentary) are not much earlier than 300 B.C., and the technological chapters lie within half a century later than that. Let us look at some of the definitions.

Duration
(Canon) Duration (*chiu*) includes all particular (different) times (*shih*).
(Exposition) Former times, the present time, the morning and the evening, are combined together to form duration.[1]

Space
(Canon) Space (*yü*) includes all the different places (*so*).
(Exposition) East, west, south and north, all are enclosed in space.[2]

Movement
(Canon) When an object is moving in space, we cannot say (in an absolute sense) whether it is coming nearer or going further away. The

[1] Cf. *SCC*, Vol IV, pt 1, p. 2. This is reminiscent of the absolutist definition of time by Strato of Lampsacus (*fl.* 300 B.C.) one of Aristotle's disciples (cf. Simplicius, *Phys.*, 789. 35), and foreshadows Newton against Leibniz.

[2] Cf *SCC*, Vol. III, p. 93. This recalls the absolutist definition of space by Strato of Lampsacus (Simplicius, *Phys.*, 618, 20) followed by Johannes Philoponus (6th century A.D.), *Phys.*, 567. 29.

	reason is given under 'spreading' (*fu*) (i.e. setting up co-ordinates by pacing).
(Exposition)	Talking about space, one cannot have in mind only some special district (*chhü*). It is merely that the first step (of a pacer) is nearer and his later steps further away. (The idea of space is like that of) duration (*chiu*). (One can select a certain point in time or space as the beginning, and reckon from it within a certain period or region, so that in this sense) it has boundaries, (but time and space are alike) without boundaries.[1]

Movement

(Canon)	Movement in space requires duration. The reason is given under 'earlier and later' (*hsien hou*).
(Exposition)	In movement, the motion (of an observer) must first be from what is nearer, and afterwards to what is further. The near and far constitute space. The earlier and later constitute duration. A person who moves in space requires duration.[2]

Thus motion, as Forke said,[3] led genetically to the idea of time as well as space. The distances left behind by a moving body, an observer, constitute space, and the changes of position of an observed moving body, such as the sun or moon, awaken the conception of time. Of course man, like all other animals and plants, had his own built-in biological clocks, sensitive to intrinsic needs and the rhythmic cycle of light and darkness.

Great debate went on among the philosophers concerning relativity and infinity. In the 4th century B.C. under the influence of Hui Shih, one of the School of Logicians (Ming Chia), many strange sayings were mooted which bear much resemblance to the Eleatic paradoxes of Greece. Hui Shih said, for example, 'The sun at noon is the sun declining, the creature born is the creature dying,' and again, 'Going to the State of Yüeh today, one arrives there yesterday.'[4] The brief moment of noon seems illusory, and if observed from different places on the earth's surface the sun is

[1] Cf. *SCC*, Vol. IV, pt. 1, p. 55.

[2] Cf. *SCC*, Vol. IV, pt. 1, p. 56.

[3] *Geschichte d. alten chinesischen Philosophie* (Hamburg, 1927), p. 413. Cf. Aristotle: 'movement is the objective seat of before-and-after-ness' (*Physica*, IV, 11, 219 a 8, 220 a 1). Aristotle was a relationist as regards time for he felt that time existed only by virtue of the motion of moving bodies. The Mohists would not have said this.

[4] Cf. *SCC*, Vol. II, pp. 190 ff.

always declining; senescence begins from the moment of conception, and indeed goes on faster the younger the organism. 'Going to Yüeh' is probably a recognition of the existence of different time-scales in different places. The Mohists held not only that time was constantly passing from one moment to another but also that particular locations in space were constantly changing; it may be that they had recognized the movement of the earth.

Space and time
(Canon) The boundaries of space (the spatial universe) are constantly shifting. The reason is given under 'extension' (*chhang*).

(Exposition) There is the south and the north in the morning, and again in the evening. Space, however, has long changed its place.

Space and time
(Canon) Spatial positions are names for that which is already past. The reason is given under 'reality' (*shih*).

(Exposition) Knowing that 'this' is no longer 'this,' and that 'this' is no longer 'here,' we still call it north and south. That is, what is already past is regarded as if it were still present. We called it south then and therefore we continue to call it south now.[1]

Perhaps the Mohists envisaged something like what we should now speak of as a universal space-time continuum within which an infinite number of local space-times coexist, and guessed that the universe would look very different to different observers according to their positions in the whole.

Then there was the infinitely long-enduring, the infinitely brief, the infinitely large and the infinitely small. Other paradoxes of the school of Hui Shih were concerned with atomism, just as those of Zeno had been. An interesting passage occurs in the *Lieh Tzu* book,[2] cast as so often in the form of a conversation between semi-imaginary characters.

[1] Cf. *SCC*, Vol. II, p. 193. This is somewhat reminiscent of the recognition by Boethus (*fl.* 50 B.C.) that motion and rest have one thing in common, the change of the independent variable of time. 'It is not correct to describe rest in a place as "place",' (cf. Simplicius, *Categ.*, 433. 30). As his predecessor Strato had said, rest is motion along the axis of time.

[2] Not finally completed till *c.* A.D. 380 but containing much material of the Warring States, Chhin and Han periods (fourth century B.C. onwards). Ch. 5, cf. *SCC*, Vol II, p. 198; translation here revised to take account of the interpretations of R. Wilhelm and A. C. Graham.

'Thang (the High King) of the Shang asked Hsia Chi saying, "In the beginning, were there already individual things?" Hsia Chi replied, "If there were no things then, how could there be any now? If later generations should pretend that there had been no things in our time, would they be right?" Thang said, "Have things then no before and no after?" To which Hsia Chi answered, "The ends and the origins of things have no limit from which they began. The origin (of one thing) may be considered the end (of another); the end (of one) may be considered the origin (of the next). Who can distinguish accurately between these cycles? What lies beyond all things, and before all events, we cannot know."

So Thang said "What about space? Are there limits to upwards and downwards, and to the eight directions?' Hsia Chi said that he did not know, but on being pressed, answered, "If they have none, there can be an infinitely (great). If they have, there must be an indivisibly (small). How can we know? If beyond infinity there were to exist a non-infinity, if within the infinitely divisible there were to exist an indivisible, then infinity would be no infinity, and the infinitely divisible would contain an indivisible. This is why I can understand the infinite and the infinitely divisible, but I cannot understand the finite and the indivisible. . . ." '

This shows the kind of arguments that were current about time and space. A little later in the same chapter Hsia Chi goes on to tell of the immense variations of the life-spans of plants and animals, pointing the moral of the relativity of time as it must seem to different living creatures. The *Chuang Tzu* book says:[1]

'Man has a real existence, but it has nothing to do with location in space; he has a real duration, but it has nothing to do with beginning or end in time.'

One could not restrict the influence or knowledge of a man to the physical space which his body happens to occupy, nor is it limited to the time-span between the moments of his conception and his death, even if these could be precisely identified. But though Hsia Chi favoured an infinite space-time continuum, the Mohists were more inclined to atomism, at least with regard to the definition of the geometrical point and the instant of time.

Instants of time
(Canon)   The beginning (*shih*) means (an instant of) time.
(Exposition) Time sometimes has duration (*chiu*) and sometimes not, for the 'beginning' point of time has no duration.[2]

[1] Ch. 23 (Legge tr. vol. 2, p. 85).
[2] Cf. *SCC*, Vol. IV, pt. 1, pp. 3 ff. There are Indian and Semitic parallels, but not historically likely to have influenced the Mohists.

In spite of this, however, atomism in the physico-chemical sense never played any role of importance in traditional Chinese scientific thinking, which was wedded to the ideas of the continuum and action at a distance.[1]

How advanced the conceptions of the Mohists and logicians were may be seen from surveys of the scientific thought of the Greeks.[2] The Mohists were very near the formulation of 'functional dependence' in the relation of motion to time. Although the Stoics, with their great emphasis on a continuous rather than an atomistic universe, developed the first beginnings of multi-valued logic and grasped one of the elements of the concept of function, the continuous variable, they could not go much further for they could not think of time as an independent variable with phenomena as its function. The description of motion by analytical geometry as change of place functionally dependent on time had to await the mathematization of physics at the Renaissance. For the Peripatetics, as we shall later see (p. 287) time was cyclical rather than linear; they could not think of time as a co-ordinate stretching to infinity from an arbitrary zero, like the abstract co-ordinates of space, in fact a geometrical dimension mathematically tractable, as Galileo did. The Mohists had no deductive geometry (though they might have developed one), and certainly no Galilean physics, but their statements often give a more modern impression than those of most of the Greeks. How it was that their school did not develop in later Chinese society is one of the great questions which only a sociology of science can answer.[3] For most of the Peripatetics moreover, there was something unreal about time,[4] and they were followed in this by most of the Neo-Platonists. In China the Buddhist schools shared this conviction as part of their general

[1] Cf. J. Needham & K. Robinson, 'Ondes et Particules dans la Pensée Scientifique Chinoise', Sciences, 1960, 1 (no. 4), 65; or alternatively SCC, Vol. IV, pt. 1, pp. 3 ff., 9 ff., 202 ff.

[2] See, for example S. Sambursky, The Physical World of the Greeks (London, 1956), pp. 181 ff., 238 ff.; also The Physical World of Late Antiquity (London, 1962), pp. 9 ff.

[3] Cf. SCC, Vol. II, pp. 165 ff., 182, 201 ff., 203.

[4] Cf. Aristotle, Phys., 217 b 33.

doctrine of the world as illusion, but the indigenous Chinese philosophers never did.

The Mohists also discussed causality. Take for instance the following:

Causation

(Canon)   A cause (*ku*) is that with the obtaining of which something becomes (comes into being, *chhêng*).

(Exposition) Causes: a minor cause is one with which something may not necessarily be so, but without which it will never be so. For example, a point in a line. A major cause is one with which something will of necessity be so (*pi jan*) (and without which it will never be so). As in the case of the act of seeing which results in sight.[1]

This distinguishes between necessary conditions and efficient causes. The former is like competence to react to a stimulus in modern biology. An enlightened understanding of the relation of causality to time is also implicit in some of the other Mohist propositions already mentioned. But this was not always so in ancient Chinese thought. Just as ancient and medieval Europeans spoke of Aristotelian final causes, so in ancient China we hear that such and such a lord, in his lifetime, was not able to obtain the hegemony of the feudal States because, after his death, human victims were sacrificed to him.[2] Both facts were felt to be part of a single pattern, not exactly timeless, but with time as one of its dimensions, in which causation could operate backwards as well as forwards. Elsewhere, within the framework of that primary natural philosophy of China which we shall discuss immediately below, there was scope for ideas of causation distinctly different from the Indian or Western atomistic picture, in which the prior impact of one thing is the cause of the motion of another. A causal event might not be strictly prior in time to its effect, bringing the

---

[1] Cf. *SCC*, Vol. II, p. 176.

[2] *Shih Chi*, ch. 5 (Chavannes tr., vol. 2, p. 45), noted by M. Granet, *Danses et Légendes de la Chine Ancienne* (Paris, 2 vols., 1926), vol. 1, pp. 104 ff. The incident is dated at 621 B.C. See below, p. 275.

latter about rather by a kind of absolutely simultaneous resonance.[1] Although this conception was quite congruent with the highly organicist tendency of Chinese science and philosophy in general, it was never very thoroughly worked out. One may doubt therefore whether in itself it played any great role in the inhibition of the rise of modern science and technology in Chinese civilisation.

The ancient Confucian school, occupied always with human affairs, was of course not interested in all these speculations, and even disapproved of them. Time entered into their considerations only in relation to the appropriate times of action of the sage in society. The 'mean' or 'norm' (chung) was the guide for emotion and action, but it must be flexible in application, for circumstances alter cases and no fixed rules of duty could be laid down, hence in the classic called the 'Doctrine of the Mean' (Chung Yung), it is a 'timely mean' (shih chung) that one must follow.[2] In the 'Book of Changes' (I Ching) too, this conception of the right timeliness in everything is very prominent. The Neo-Confucian school of the Middle Ages, however, had an altogether different approach. These scholastic thinkers were by that time (eleventh to thirteenth centuries A.D.) aware not only of all the Mohist and Taoist speculations of old, but also of the numerous philosophies of Buddhism,

---

[1] On these unfamiliar views of causality in Chinese natural philosophy, for example 'reticulate causation' and 'synchronistic causation', see SCC, Vol. II, pp. 288 ff., and the following works of C. G. Jung: Wilhelm, R. & Jung, C. G., The Secret of the Golden Flower; a Chinese Book of Life, including a translation of the Thai I Chin Hua Tsung Chih, Eng. tr. by C. F. Baynes (London, 1931), esp. p. 142; Jung, C. G. & Pauli, W., Naturerklärung und Psyche (Rascher, Zürich, 1952, Studien aus dem Jung Institut, no. 4), Eng. tr. by R. F. C. Hull (London, 1955), containing 'Synchronicity, an Acausal Connecting Principle' by Jung, and 'The Influence of Archetypal Ideas on the Scientific Theories of Kepler' by Pauli; Jung, C. G., 'Über Synchronizität', Eranos Jahrbuch, 1952, 20, 271; Jung, C. G., 'Synchronicity; an Acausal Connecting Principle', in The Structure and Dynamics of the Psyche (London, 1960, Collected Works, vol. 8). For a discussion of similar ideas from a quite different standpoint read N. R. Hanson, 'Causal Chains', Mind, 1955, 64 289.

[2] Chung Yung, II, ii. Cf. Fêng Yu-Lan, History of Chinese Philosophy (London, 1937), vol. I, pp. 371, 391. This idea is something like the idios kairos (ἴδιος καιρός) of the early Christian writers, the appropriate or decisive moment for action, divine or human.

and they adopted eclectically whatever suited the purpose of their new synthesis. It would not be possible here to examine in detail their various attitudes to time, but most of them accepted it as real, objective and infinite. So, for example, Shao Yung (A.D. 1011 to 1077).[1] Some, however, such as his son Shao Po-Wên (A.D. 1057 to 1137) regarded time as subjective, since for the eternal Tao there could be no past, present or future.[2] Most, as we shall see, believed in cycles of recurrence within time.[3]

In this they were drawing on ancient Taoism. Nothing could be more striking than the appreciation of cyclical change, the cycle-mindedness, of the Taoists. 'Returning is the characteristic movement of the Tao (the Order of Nature)' says the *Tao Tê Ching*.[4] 'Time's typical virtue', wrote Granet, 'is to proceed by revolution.'[5] Indeed time (*shih*) is itself generated, some thought, by this uncreated and spontaneous (*tzŭ-jan*) never-ceasing circulation (*yün*).[6] The whole of Nature (*thien*), the Taoists felt, could be analogized with the life-cycles of living organisms. 'A time to be born and a time to die,' a time for the founding of a dynasty and a time for its supersession. This was the meaning of destiny (*ming*), hence the expressions *shih-yün* and *shih-ming*. The sage accepts; he knows not only how to come forward but also how to retire.[7] This preoccupation with cycles had interesting results in later times when Chinese scientific thought appreciated the existence of natural cycles, sometimes before other civilizations; for example

---

[1] Cf. *SCC*, Vol. II, pp. 455 ff.

[2] Cf. A. Forke, *Geschichte d. neueren chinesischen Philosophie* (Hamburg, 1938), p. 42.

[3] One of the Ming scholars, Tung Ku, replied to a questioner that time could be said to have a beginning if you were speaking of a single world-period (*yuan*), but not if you were speaking of the endless chain of all the world periods. Cf. *SCC*, Vol. III, p. 406.

[4] Ch. 40; cf. *SCC*, Vol. II, pp. 75 ff.

[5] *La Pensée Chinoise* (Paris, 1934), p. 90.

[6] I take this formulation from a paper by Fukunaga Mitsuji and the commentary of Than Chieh-Fu on the Mohist proposition quoted on p. 220 above. It has Peripatetic and Neo-Platonic parallels; cf. Sambursky, *Physical World of Late Antiquity*, p. 15.

[7] Cf. *SCC*, Vol. II p. 283.

the meteorological water cycle,[1] or the circulation of the blood and *pneuma* in the human and animal body.[2] It was prevalent in nearly all the schools, not only the Neo-Confucians, and in this case as in others a close relation may be perceived between the cyclical world-view and that other paradigm of Chinese scientific naturalism, wave-theory as opposed to atomism.[3]

An important question here is how far the individual cycles, or particular parts of cycles, were compartmentalized, separated off from one another into discrete units, for the ancient and medieval Chinese thinkers. In an influential book, Granet concluded that time in ancient Chinese conception was always divided into separate spans, stretches, blocks or boxes, like the organic differentiation of space into particular expanses and domains.[4] *Shih* (*time*) always seemed to imply specific circumstances, specific duties and opportunities;[5] it was essentially discontinuous 'packaged' time.[6] This conclusion was based, not on the study of the

---

[1] Cf. *SCC*, Vol. III, pp. 467 ff.

[2] This subject will be fully treated of in *SCC*, vol. VI; meanwhile see P. Huard & Huang Kuang-Ming (M. Wong), 'La Notion de Cercle et la Science Chinoise', *Archives Internat. d'Hist. des Sciences*, 1956, **9** 111. A remarkable result of this circulation-mindedness has recently been brought to light by Lu Gwei-Djen & J. Needham, 'Medieval Preparations of Urinary Steroid Hormones', *Nature*, 1963, **200**, 1047. In Renaissance Europe the philosophy and mysticism of the circle and circulation were immensely important in the developing phases of modern science. This influence has been traced in many papers by W. Pagel, among which we can only mention here 'Giordano Bruno; the Philosophy of Circles and the Circular Movement of the Blood', *Journ. Hist. Med. & Allied Sci.*, 1951, **6**, 116.

[3] Cf. *SCC*, Vol. IV, pt. 1, pp. 3 ff., 9 ff.

[4] Granet, *La Pensée Chinoise*; the Chinese 'preferred to see in time an ensemble of eras, seasons and epochs' (p. 86); 'Time and space were never conceived apart from concrete actions' (p. 88); the Chinese 'decomposed all time into periods just as they decomposed all space into regions' (p. 96); the Chinese 'never bothered about imagining time and space as homogeneous matrices suitable for housing abstract concepts' (p. 113).

[5] Granet, loc. cit., p. 89.

[6] Granet often used to call this 'liturgical time' because of its connection both with the ceremonies of the imperial cosmic religion and with those which marked the incidents of lives within the individual family. Much importance was attached to the correct vesting of the emperor and his attendants in accordance with the season when performing the rites of the Ming Thang or Cosmic Temple. Full details will be found in W. E. Soothill, *The Hall of Light, a Study in Early Chinese Kinship*

philosophical schools already mentioned, but (as was Granet's way) on the mythology, folk-lore and general world-outlook of the classics and other ancient writings, including Chhin and Han literature. This world-outlook was systematized by yet another group of thinkers, the proto-scientific School of Naturalists (Yin Yang Chia), headed by Tsou Yen (*fl.* 350 to 270 B.C.), oldest of the Chi-Hsia Academicians.[1] The Naturalists elaborated the theory of the two fundamental natural forces, Yin and Yang, the theory of the five elements, and the system of the symbolic correlations, in which a great number of objects and entities were classified by fives in correspondence with the elements Wood, Fire, Earth, Metal and Water.[2] Since the seasons were prominent in this classification,[3] as also the double-hours of the day and night (because of the use of the twelve cyclical signs to denote them), time was to that extent 'boxed'; and since this differentiation of time was duly extended to states, dynasties, rulers and reign-periods, one can understand the exceptional political power of Tsou Yen and his followers in the late feudal and early imperial period. Success or failure in peace and war might well hinge, it was thought, on adherence to the appropriate element and all its corresponding entities in the symbolic correlation system; hence

---

(London, 1951), pp. 30 ff. Once Summer had officially begun it would have been an unthinkable affront to Heaven and Earth to wear the green robes appropriate to Wood, the element of Spring; all vestments, banners and cult objects had to be changed to red, the colour appropriate to Fire, the element of Summer. Such practices are familiar to us in the form of the liturgical colours of Western Catholic worship, though these do not symbolize 'blocks' of time as the ancient Chinese ones did, nor are they charged with superstitious anxiety, fear of forfeiting Heaven's favour. Then there were also the ritual lapses of time in family customs, the 'rites de passage' which had to be accomplished in each individual's life-history, the temporary prohibitions, the periodical festivals, the duties performed at fixed intervals. See Granet, *Pensée Chinoise*, pp. 97 ff. and also a famous study of his, 'Le Dépot de l'Enfant sur le Sol', published originally in *Revue Archéologique*, 1922, (5e sér.) **14**, 10, and reprinted in *Etudes Sociologiques sur la Chine* (Paris) 1953, p. 159.

[1] See *SCC*, Vol. II, pp. 232 ff.

[2] Cf. *SCC*, Vol. II, pp. 242 ff., 253 ff., 261 ff., 273 ff.

[3] The four seasons were correlated with four of the elements. The sixth month, however, was placed under the aegis of Earth, thus making the seasons up to the right number, five.

the prestige of the proto-scientific prognosticators. All this formed
the primary Chinese world-view on which was based the later
traditional natural philosophy of alchemists and acoustic experts,
geomancers and pharmacists, smiths, weavers and master-
craftsmen throughout the centuries. In my own description of it I
wrote, 'For the ancient Chinese, time was not an abstract parameter, a succession of homogeneous moments, but was divided
into concrete separate seasons and their subdivisions.'[1] The idea of
succession as such was subordinated to that of alternation and
interdependence.[2]

This was assuredly true but it was not the whole story.[3] First,
although the theories of the Naturalists were in general widely
accepted, and even to some extent developed for proto-scientific
purposes, they were much less powerful in some realms of Chinese
society than in others. The competing schools of Mohists and
Logicians were of course never interested in them, and they played
relatively little part in the long evolution of astronomy and cosmology; furthermore, as we shall see, the historians on the one
hand and the mass of the people on the other, when they engaged
in long-term sociological study and speculation, found no use for
the compartmentalized time of the Naturalists. And secondly
while cyclical recurrence was indeed prominent in the natural
philosophy, it was almost entirely the cycles of the annual seasons,
months, days, hours, etc., and of those which present themselves
in biological or social organisms—long-term astronomical

[1] SCC, Vol. II, p. 288.
[2] Granet, La Pensée Chinoise, pp. 329 ff.; cf. SCC, Vol. II, pp. 289 ff.
[3] It is a little unfortunate that Western writers on time have assumed that these
ideas can fully represent Chinese thinking. Thus G. J. Whitrow, Natural Philosophy
of Time (London and Edinburgh, 1961), p. 58, has concluded that discontinuous
'packaged' time was the only time the Chinese knew of. He kindly acknowledges
the help of SCC, Vol. II, but overlooked footnote (f) on p. 288 which would have
directed him to another Chinese world, that of the Mohists. Of course he himself
was devoting only a single footnote to China. One could hardly compress all the
European ideas about time held through the ages into a single footnote, and few
Chinese writers, I am sure, would attempt to do so. Western indifference to Chinese
thought has been such that Whitrow deserves praise, not blame, for his effort—yet
there may still be a lesson in this contrast.

periods played an insignificant part, and 'Great Year' conceptions (see p. 288), with their consequence of temporal recurrence, none at all. Thirdly, it may be noted that the political applications of the natural philosophy were more and more doubted as time went on. In the Chhin and Han periods there were intense and anxious debates about the proper colour, musical notes and instruments, sacrifices, etc., appropriate to a particular dynasty or emperor, and such questions could still be a live issue in the sixth century A.D.,[1] but after that the political significance of the symbolic correlations seems to have played a steadily decreasing part in men's minds.[2] In sum there was both compartmentalized time and continuous time in Chinese thinking.[3] Both were important in different ways, the former for some of the sciences and technology, the latter for history and sociology.[4]

Here we may be putting a finger on one of the keys for the answer to the question why modern science did not develop spontaneously in China. In so far as the traditional natural philosophy was committed to thinking of time in separate compartments or boxes perhaps it was more difficult for a Galileo to arise who should uniformize time into an abstract geometrical coordinate, a continuous dimension amenable to mathematical handling. But then Chinese astronomy does not prominently show this compartmentalization of time—for example one never finds a particular planetary revolution associated with an element or a colour, though the planets themselves were of course named after the elements. And in the story of the associated invention of

---

[1] About A.D. 543 the great Taoist swordsmith and metallurgist Chhiwu Huai-Wên, probably the inventor of co-fusion steel, advised the emperor Kao Tsu of the Eastern Wei to change the colour of that dynasty's flags from red to yellow in accordance with five-element theory, in order to conquer the Western Wei.

[2] This if fully worked out might add another chapter to the history of the spontaneous development of scientific criticism in China. Of rationalist scepticism there was never any lack (cf. SCC, Vol. II, pp. 365 ff.), and there are other examples of the development of critical scientific thought quite independently of European Renaissance influences (see SCC, Vol. IV, pt. 1, pp. 189 ff.).

[3] Later on in Chinese history, as we shall see, the Indian ideas of long-term recurrence had considerable success in China, so there was cyclical time too.

[4] And proto-archaeology also, as we shall see, below.

mechanical clockwork (entirely Chinese, as we shall see), there is nothing to indicate inhibition by any ideas of sharp boundaries between stretches of time; the clocks were tended continually, and they ticked away continually decade after decade

Truly, the cyclical does not necessarily imply either the repetitive or the serially discontinuous. The cycle of the seasons in the individual year (*annus*) was but one link (*annulus*) in an infinite chain of duration, past, present and future. By the use of two interlocking sets of cyclical characters, one of ten and one of twelve, the Chinese were also accustomed to measure time in sexagenary periods. These were used for the day-count from the fifteenth century B.C. onwards, and for the year-count from the first century B.C., thus giving a system independent of celestial phenomena, with a 'week' (*hsün*) of ten days. In a civilization primarily agrarian the people had to know exactly what to do at particular times, and so it came about that the promulgation of the luni-solar calendar (*li*) in China was the numinous cosmic duty of the imperial ruler (the Son of Heaven, Thien Tzu). Acceptance of the calendar was the demonstration of fealty, somewhat analogous to the authority of the ruler's image and superscription on the coinage in other civilizations.[1] Since celestial magnitudes are incommensurable and subject to slow saecular change, continual work on the calendar was necessary through the ages, and few were the mathematicians and astronomers in Chinese history who did not work upon it. Between 370 B.C. and A.D. 1742 no less than 100 'calendars' or sets of astronomical tables were produced, embodying constants of ever greater accuracy, and dealing with the determination of solstices, day, month and year lengths, the motions of sun and moon, planetary revolution periods, and the like.[2] Metonic, Callippic, and saros-like eclipse cycles were early

---

[1] Cf. Granet, *La Pensée Chinoise*, p. 97.

[2] Something on this important subject will be found in *SCC*, Vol. III, pp. 390 ff. but I regret that our treatment of it was not really adequate. We felt at that time that the interest of the calendar was rather archaeological and social-historical than scientific, not fully appreciating that each 'calendar' had been a full set of astronomical tables and constants intended always to constitute a substantial improvement over all its predecessors. One difficulty (even for those familiar with the

recognized.[1] The calendar had indeed a central role in the history of Chinese science and culture. And since one calendar system covered many years or decades, it blended into history itself.

TIME, CHRONOLOGY AND CHINESE HISTORIOGRAPHY

The closeness of the relation between the two can be seen strikingly in the title of a high official, the Thai Shih (or Thai Shih Kung, or Thai Shih Ling). Today we translate this, from all medieval and late texts, as 'Astronomer-Royal', for he was the head of the Bureau of Astronomy in the civil service, and less and less occupied with State astrology as time went on. In the early Han however, 'Astrologer-Royal' would not be a wrong rendering, but what is really significant is that 'Historiographer-Royal' would not be wrong either. This was in fact the title borne by the first of China's great historians, Ssuma Chhien, and by his father Ssuma Than before him. Someone has suggested that perhaps the term 'Chronologer-Royal' would be the best, for the office was certainly thought of as combining an earthly archivistic with a heavenly uranographic function. Traced back etymologically as far as possible, the character *shih* is a pictograph showing a hand holding something, perhaps a sign for centralness, which may have meant impartiality, or location at the capital; perhaps the cup for counting the tallies held up by the referee of the ancient archery contests—a servitor therefore, but impartial and intelligent. Whether the primary function of the Thai Shih was terrestrial or celestial is not yet decided;[2] certain at least it is that

Chinese language) is that the greatest authority on the history of Chinese ephemerides, Dr Yabuuchi Kiyoshi, has published his voluminous work almost entirely in Japanese. It would be a great service to learning if this could be digested into a practical handbook either in Chinese or a Western language. Meanwhile the monograph of Chu Wên-Hsin, *Li Fa Thung Chih* (Shanghai, 1934), on the Chinese calendar-systems remains indispensable.

[1] Cf. *SCC*, Vol. III, pp. 406 ff.

[2] Among recent interesting discussions of the problem we may mention F. Jäger, 'Der heutige Stand Schï-ki [*Shih Chi*] Forschung', *Asia Major*, 1933, 9, 25; B. Watson, *Ssuma Chhien, Grand Historian of China* (New York, 1958), pp. 70 ff., 204 ff., 220; F. A. Kierman, *Ssuma Chhien's Historiographical Attitude as Reflected in*

by the middle of the first century A.D. the Bureau of Astronomy and the Bureau of Historiography were two entirely separate branches of the bureaucratic organization. The second of these was charged with conserving the records and archives of the current dynasty, and with writing the history of the preceding one, and in theory it was free from manipulation by the reigning monarch or the officials of the day.[1]

It was quite characteristic, in view of Chinese realism about time, that China should have possessed perhaps the greatest of all ancient historical traditions.[2] One can say without hesitation that the Chinese were the most historically-minded of all ancient peoples; a quality which makes the dating of events in their civilization comparatively easy. Archaeology too proves this daily, for objects and inscriptions were meticulously dated. No other culture has given us so great a mass of historical writing as that constituted by the twenty-five official dynastic histories, starting with the *Shih Chi* (Historical Records) of Ssuma Chhien already mentioned, which was finished about 90

Four Late Warring States Biographies (Wiesbaden, 1962) pp. 4 ff., 48 ff. There is evidence that the office of Thai Shih could be, and often was, hereditary, for Ssuma Than told his son that their family came of a line of Thai Shih of the imperial house of Chou. Thaishih could also be derivatively a surname, as in the case of Thaishih Chiao of Chhi (*fl.* 320 to 270 B.C., see Kierman, ibid., p. 39), and Thaishih Shu-Ming (A.D. 474 to 546) a Taoist scholar and mutationist.

[1] And to a large extent, in practice. For an account of this system see Yang Lien-Shêng, 'The Organisation of Chinese Official Historiography; Principles and Methods of the Standard Histories from the Thang through the Ming Dynasty', art. in *Historians of China and Japan*, ed. W. G. Beasley & E. G. Pulleyblank London, 1961), p. 44. The contribution of A. F. Hulsewé in the same volume (p. 31), 'Notes on the Historiography of the Han Period', shows how it developed from the ancient Thai Shih organization of astronomer-annalists. By the Thang period it had reached its definitive form.

[2] On Chinese historiographic traditions in general, see the introduction to the book just mentioned (by E. G. Pulleyblank & W. G. Beasley), and especially the small but now long classical treatise of C. S. Gardner, *Chinese Traditional Historiography* (Cambridge, Mass. 1938, repr. 1961). Han Yu-Shan's *Elements of Chinese Historiography* (Hollywood, Calif. 1955) is also a useful reference work, though with many inaccuracies. For those who can read Chinese there is an up-to-date monograph by Chin Yü-Fu, *Chung-Kuo Shih Hsüeh Shih* (A History of Chinese History-Writing), (Peking, 1962). A brief survey of Chinese historiography is given in *SCC*, Vol. I, pp. 74 ff.

B.C., and ending with the *Ming Shih* (History of the Ming Dynasty), completed in A.D. 1736. That customary expression 'the Chinese annals' so often encountered, shows how little Western writers have understood of the pattern of the dynastic histories. To be sure, these contain the basic annals of the successive reigns, but they also contain a great quantity of treatises on special subjects such as astronomy, economics, civil service organization, administrative geography, hydraulic engineering, taxation and currency, law and justice, court ceremonial, etc., etc. Lastly they embody a vast wealth of biographies of individuals, and this is among the most valuable of all the material in the dynastic histories. Fortunately, the greatness of the Chinese historical tradition, the work of a people who took time seriously, is now more and more appreciated by Western scholars.[1]

The question has been raised, however, whether the time of Chinese historians was not 'boxed time' rather than continuous time.[2] It is quite true that the idea of a single era count, such as the Olympic dating from 776 B.C., or the Seleucid from 311 B.C., or our own Christian era defined in the early sixth century A.D., did not spontaneously originate in China.[3] Years were counted in terms of dynasties and reigns, and (from about 165 B.C. onwards) special regnal periods (*nien hao*) within reigns. But the historians worked out a coherent 'single track' theory of dynastic legitimacy and made great efforts to correlate the chronology of events in concurrent minor dynasties, overlapping kingdoms and barbarian peoples, with the graduations on the main time-scale adopted.

---

[1] By those at least who have taken the trouble to acquaint themselves with China's historiography. A regrettable example of ignorant judgment is cited by E. G. Pulleyblank (in W. G. Beasley & E. G. Pulleyblank, op. cit.), p. 135, and to this, I fear, must be added the lecture by H. Butterfield, *History and Man's Attitude to the Past; their Role in the Story of Civilisation* (Foundation Day Lecture, London School of Oriental Studies, 1961).

[2] In an article by O. van der Sprenkel, 'Chronology, Dynastic Legitimacy, and Chinese Historiography', contributed to the Study Conference at the London School of Oriental Studies in 1956 and circulated at that time in mimeographed form, but unfortunately not printed in *Historians of China and Japan*, ed. W. G. Beasley & E. G. Pulleyblank, 1961.

[3] Cf. footnote on p. 286 below.

FIGURE 30 A late Chhing illustration of an event recorded in the *Shu Ching*
(Historical Classic) *c.* 9th century B.C.; the taking over of the official civil service
of the defeated Shang dynasty by the new dynasty of the Chou. The caption says:
'Chou Kung (the Duke of Chou) makes an announcement to the officials of the
(displaced dynasty of) Shang', (To Shih ch. from *Shu Ching Thu Shuo*).

One of the greatest astronomers who thus joined Joseph Scaliger[1] and Isaac Newton[2] in the field of chronology was Liu Hsi-Sou (*fl.* A.D. 1060) whose *Liu Shih Chi Li* (Mr Liu's Harmonized Calendars) embodied the results of his *chhang shu* (art of reconciling long-period data) by means of the sexagenary cycles, identification of intercalary months, solstice dates, etc.[3]

Furthermore Chinese historiography was by no means confined to the set framework of the dynastic history, for as time went on there grew up various forms of 'continuity history-writing' which dealt with long periods of time, including the rise and fall of several dynasties. Ssuma Chhien himself had set a pattern for this, since his *Shih Chi* began with the remotest antiquity and came down to about 100 B.C. in the Earlier Han dynasty, but he did not theorize much about the work of the historian. The philosophy of history was brilliantly studied, however, in the Thang period with the *Shih Thung* (Generalities on History) of Liu Chih-Chi (A.D. 661 to 721), finished in A.D. 710—the first treatise on historiographi-

---

[1] Joseph J. Scaliger (A.D. 1540 to 1609) founded modern historical chronology with his *Opus Novum de Emendatione Temporum* (*Thesaurus Temporum*), Paris, 1583. See J. W. Thompson & B. J. Holm, *A History of Historical Writing*, 2 vols. (New York, 1942), vol. 2, p. 5. Although this work is invaluable for the Western world, and made a creditable attempt to say something about Arabic, Persian and Mongol historians, it deliberately if tacitly excluded China from its survey (hence the stricture of Pulleyblank just mentioned), not hesitating however on that account to affirm in its introduction the pre-eminent historical-mindedness of Christian Europe.

[2] Sir Isaac Newton, *The Chronology of Ancient Kingdoms Amended, to which is prefixed a Short Chronicle from the First Memory of Things in Europe to the Conquest of Persia by Alexander the Great* (London, 1728). See on this F. Manuel, *Isaac Newton, Historian* (Cambridge, 1964).

[3] For the general background of Liu Hsi-Sou see Yabuuchi Kiyoshi, 'The Development of the Sciences in China from the 4th to the end of the 12th Century A.D.', *Journ. World History*, 1958, **4**, 330. *Chhang shu* methods had first been used by Tu Yü in the third century in his studies on the *Chhun Chhiu* period. Liu Hsi-Sou's work, which covered the time from the beginning of the Han to the end of the Wu Tai periods, was extended to the Sung, Yuan and Ming by a great successor, Chhien Ta-Hsin (A.D. 1728 to 1804). We now possess, of course, chronological tables for Chinese history of great detail and precision, for example A. C. Moule & W. P. Yetts, *The Rulers of China, 221 B.C. to A.D. 1949* (London, 1957).

cal method in any language,[1] quite worthy of comparison with the work of the European pioneers Bodin and de la Popelinière eight and a half centuries later.[2] At that time China was also to have her Giambattisto Vico, Chang Hsüeh-Chhêng (A.D. 1738 to 1801).[3] It was Liu Chih-Chi's own son, Liu Chih (*fl. c.* A.D. 732), and another Thang scholar Tu Yu (A.D. 735 to 812) who invented a new form of encyclopedic institutional history, the former with his *Chêng Tien* (Governmental Institutes) the latter with the famous *Thung Tien* (Comprehensive Institutes; a Reservoir of Source Material on Political and Social History) of A.D. 801. But the climax of this type of work was not reached until the Yuan period, when in 1322 the *Wên Hsien Thung Khao* (Comprehensive Study of the History of Civilization), by Ma Tuan-Lin, saw the light.[4] His lucid and outstanding treatise, in 348 chapters, was essentially a general history of institutions, which, together with the social structures and economic situations implied by them, seemed to Ma a much more important form of history than any chronological catalogue of contingent events. This search for causal sequences in history more fundamental than dynastic and military mutations and permutations was remarkably advanced for its time; indeed it paralleled the sociological history initiated by

[1] On this see the excellent study of E. G. Pulleyblank, 'Chinese Historical Criticism; Liu Chih-Chi and Ssuma Kuang', art. in *Historians of China and Japan* ed. W. G. Beasley & E. G. Pulleyblank, p. 135.

[2] Jean Bodin (A.D. 1520 to 1596), *Methodus ad facilem Historiarum Cognitionem* (Paris, 1566), and L. V. de la Popelinière (A.D. 1540 to 1608), *Histoire des Histoires; Premier Livre de l'Idèe de l'Histoire Accomplie* (Paris, 1599). These were the first Western books to discuss the laws of historical causation and development and to lay the foundation for a method and critique of history; cf. Thompson & Holm, loc. cit., vol. 1, pp. 561, 563, vol. 2, p. 5.

[3] On him see the special study of P. Demiéville, 'Chang Hsüeh-Chhêng and his Historiography', art. in *Historians of China and Japan*, ed. W. G. Beasley & E. G. Pulleyblank, p. 167, esp. p. 184. On Vico (A.D. 1668 to 1744) see Thompson & Holm, op. cit., vol. 2, pp. 92 ff.

[4] On these three great undertakings see, besides the paper of E. G. Pulleyblank on Liu Chih-Chi, E. Balazs' 'L'Histoire comme Guide de la Pratique Bureaucratique; les Monographies, les Encyclopédies, les Recueils de Statuts', art. in *Historians of China and Japan*, ed. W. G. Beasley & E. G. Pulleyblank, p. 78. Also Han Yu-Shan, op. cit., pp. 60 ff.

Ma's near contemporary the great Ibn Khaldūn,[1] and the history of institutions later to be achieved by Pasquier, Giannone and de Montesquieu.[2] The *Wên Hsien Thung Khao* also included elaborate critical analyses of the original sources, and is informed throughout by remarkable perspicacity and profound judgment.

The first move for narrative continuity history came early in the sixth century A.D., when emperor Wu of the Liang (Hsiao Yen) commissioned Wu Chün to produce a *Thung Shih* (General History); this he did, in 620 chapters, but it has not come down to us.[3] Still preserved, however, is the *Thung Chih* (Historical Collections) of Chêng Chhiao (A.D. 1104 to 1162), c. A.D. 1150. Chêng Chhiao was more successful in his theory of 'synthesis' or 'inter-relatedness' (*hui thung chih tao*) than in his practice as a historian, and the Lüeh or monograph section of his work, a topically arranged historical encyclopedia, is the only part now used and admired.[4] He had indeed been preceded by the grandest of all the Chinese continuity histories, the *Tzu Chih Thung Chien* (Comprehensive Mirror (of History) for Aid in Government), finished in A.D. 1084 by Ssuma Kuang (A.D. 1019 to 1086) and a team of collaborators.[5] It covered the whole period from 403 B.C. to A.D. 959 in

[1] As is now well known, 'Abd al-Ramān ibn Khaldūn (A.D. 1332 to 1406) worked out a general theory of historical development, embodying climate, geography, moral and spiritual forces, and laws of natic.al progress and decay—sociological history in fact—in his *Kitāb al-'Ibar wa-Diwān al-Mubtada' wa-l-Khabar-fi Ayyām al-'Arab wa-l-Ajam wa-l-Barbar* (Book of Instructive Examples and Register of Subjects and Predicates dealing with the History of the Arabs, the Persians and the Berbers).

[2] E. Pasquier (A.D. 1529 to 1615), *Les Recherches de la France*, Paris, 1560, 1611; Pietro Giannone (A.D. 1676 to 1748), *Storia Civile del Regno de Napoli*, Naples, 1723; Louis de Secondat, de Montesquieu (A.D. 1689 to 1755), *L'Esprit des Lois*, Paris, 1748. On these see Thompson & Holm, loc. cit., vol. i, p. 6, vol. 2, pp. 61 ff., 90, 561. Butterfield, op. cit., p. 13, wondered whether any non-European civilization had developed the history of laws and institutions.

[3] Han Yu-Shan, op. cit., p. 49.

[4] Han. loc. cit., pp. 49, 61, and Balazs, op. cit., pp. 84, 90.

[5] Balazs, op. cit., Yang Lien-Shêng, op. cit., and Pulleyblank, op. cit. The latter article gives a vivid account of the methods of Ssuma Kuang and his collaborators Liu Pin, Liu Shu and Fan Tsu-Yü. Some of the techniques of the medieval Chinese historians were remarkably modern. Thus for example in the Sung Bureau of Historiography coloured inks of various kinds were used to distinguish the different

239

354 chapters, and during the following centuries was constantly commented upon, abridged, digested, imitated and extended. Its thirteen centuries' breadth of canvas invites comparison with the fourteen centuries of Edward Gibbon.[1] It generated yet another form of history, the *chi shih pên mo* (narratives of major sequences of events from beginning to end), for about A.D. 1190 Yuan Shu (*fl.* A.D. 1165 to 1205), oppressed by the mass of material in the 'Comprehensive Mirror', selected 239 particular topics and followed each through separately in his *Thung Chien Chi Shih Pên Mo*.[2] This again gave rise to a whole genre of historical writing in subsequent centuries. Thus did the Chinese overcome the 'compartmentalization' of time.

It is worth while to notice particularly here the title of Ssuma Kuang's masterpiece—'for aid in government'. There was a seeming paradox in the Chinese conception of history. In China good history was considered (*a*) objective, (*b*) official, and (*c*) normative. Confucius himself, in his doctrine of the 'rectification of names' (*chêng ming*), had insisted on a spade being called a spade, no matter how powerful the interests which wanted it to be called a shovel;[3] and it was the duty of the historian, even though he was what we should call a civil servant, an eater of the bread of authority, to render judgments on the acts of the past without fear or

texts, a practice going back as far as about A.D. 500 when the great pharmaceutical naturalist Thao Hung-Ching used it in editing the pharmacopoeia. Then the historian Li Tao (A.D. 1115 to 1184) was renowned for his elaborate filing system, a row of ten cabinets each containing twenty drawers, all notes and documents relating to a particular year being filed in one of the drawers, and eventually sorted into folders in chronological order according to month and day. This system was used in the production of his *Hsü Tzu Chih Thung Chien Chhang Pien* (Supplementary Continuation of the Comprehensive Mirror of History, for Aid in Government), bringing the narrative down to A.D. 1180. How far such filing systems originated with Li Tao is not known, but it is more than probable that they had been developing ever since the days of Wu Chün (Thao Hung-Ching's contemporary), and Ssuma Kuang and his collaborators must surely have used something of the kind. On the life and work of Li Tao see Sudo Yoshiyuki in *Komazawa Shigaku* 1957, **6**, 1.

[1] Cf. Thompson & Holm, op. cit., vol. 2, pp. 74 ff.

[2] On this see Pulleyblank, loc. cit., p. 158.

[3] See Hsü Shih-Lien, *The Political Philosophy of Confucianism* (London, 1932), pp. 43 ff.; Yang Lien-Shêng, loc. cit. p. 52.

favour, 'for the punishment of evil-doers, and the praise of them that do well.'[1] Government in China could and did bestow titles and honours on the dead as well as the living (again illustrating an attitude somewhat different from our own to the dimension of time), so it was natural that the making of a just and definitive record of the past should be its function also. Finally history served an essential moral purpose 'for aid in government', for guiding administrative action, encouraging virtue and deterring vice. Such was the basic 'praise-and-blame' (*pao pien*) theory of Chinese historiography, a high endeavour of the human spirit, however displeasing to the Tory historians of the modern West.[2] Anything apparently paradoxical in this combination was resolved by a profound if tacit conviction which ran through all generations of Chinese historical writers, namely that the process of social unfolding and development had an intrinsic logic, an indwelling Tao, which rewarded 'human-heartedness' (*shan hsing, pu hu jen chih hsin, tshê yin chih hsin*)[3] with good social consequences in the long run and when all balances were struck, while its opposite brought irretrievable evil.[4] This induction was felt to have over-

---

[1] There has been much discussion of the objectivity and reliability of the Chinese official historians (apparently unknown to critics such as Butterfield) and in general sinologists have reached very favourable conclusions. One can mention H. H. Dubs, 'The Reliability of Chinese Histories', *Far Eastern Quarterly*, 1946, **6**, 23; E. R. Hughes, 'Importance and Reliability of the I Wên Chih', *Mélanges Chinois et Bouddhiques*, 1939, **6**, 173; and the debate in *Oriens Extremus*, H. H. Frankel, 'Objectivität und Parteilichkeit in d. off. Chin. Geschichtsschreibung', 1958, **5**, 133, with the reply by H. H. Dubs, 1960, **7**, 120. For a warm appreciation of the ideals of Chinese historians see E. Haenisch, 'Der Ethos d. Chin. Geschichtsschreibung', *Saeculum*, 1950, **1**, 111. The reliability of the voluminous astronomical records in the Chinese official histories is a separate question, and has been discussed in *SSC*, Vol. III, pp. 417 ff.

[2] I am thinking of course of H. Butterfield's stimulating little book, *The Whig Interpretation of History* (London, 1951).

[3] Cf. Mencius, II (1), vi, 1–7 and VI (1), vi, 4–7.

[4] This was quite different from the *karma* of the Buddhists, for retribution did not necessarily overtake the individual either in this life or in some other, but it did bring about the ruin of his house or family or dynasty or social group in the end. This Confucian form of 'cosmic reciprocity' was essentially social, and in origin it antedated by many centuries the entry into China of Buddhism and Indian ideas of reincarnation. One can find an almost epigrammatic statement of it in the *Huai Nan Tzu* book (ch. 13, Morgan tr., p. 160), *c*. 120 B.C.

whelming empirical justification. Thus history is the manifestation of the Tao, and has its origins in Heaven.[1] How could anyone ever have imagined that the time-sense of the Chinese was inferior to that of Europeans? Almost the opposite could be said; for the incarnation of the Tao in history was a continuous process, ever renewed.[2]

It would really be true to say that in Chinese culture, history was the 'queen of the sciences', not theology or metaphysics of any kind, never physics or mathematics. History thus even helped to inhibit the growth of the sciences of Nature, confined as they were to hypotheses of medieval type down to the end of their auto-chthonous development, and never achieving that mathematiza-tion which generated modern science in Europe and in Europe only. Some indeed have gone so far as to urge that the pre-eminence of history is almost alone sufficient to account for the failure of Chinese culture to develop systematic logic on Aristo-telian and scholastic lines out of the brilliant beginnings of the Mohists and Logicians.[3] In so far as explicit syllogistic logic would have helped the growth of the sciences (a somewhat debatable point),[4] here was another limiting factor, for it was not available in medieval China. Seeking for the concrete causes of this great

[1] According to a saying of Confucius current in Han times, 'there is more pun-gency and clarity in showing the Tao in action, in the facts themselves, than in expressing the Tao in empty words', (*Shih Chi*, ch. 130). The Tao inheres in Nature and history, it cannot be looked for outside the world.

[2] No one was more explicit about this than the great Chang Hsüeh-Chhêng (see the exposition of Demiéville, loc. cit., pp. 178 ff.). The canonical classics, he said, were really history—there was no distinction between *ching* and *shih*—and by the same token all history had canonical value. Chang Hsüeh-Chheng in fact canonized history, thus becoming the predecessor of his younger contemporary Hegel and in the next generation Karl Marx, although of course completely unknown to them.

[3] These were perhaps almost too brilliant, since they show many traces of dia-. lectical, as opposed to formal, logic, and this tendency was powerfully reinforced by the Indian dialectical logic introduced with Buddhism (cf. *SCC*, Vol. II, pp. 77, 103, 180 ff., 194, 199, 258, 423 ff., 458). The natural sciences could hardly benefit by dialectical logic without having passed through the stage of formal logic.

[4] It is not clear whether it helped more than it hindered. In any case, the mystical-empirical factor may have been much more important than the logical in the European scientific revolution (cf. *SCC*, Vol. II, pp. 89 ff., 200 ff.), indeed as much so as the mathematical.

difference, Stange has contrasted the social conditions of the ancient philosophers of China and Greece, pointing especially to the proto-feudal bureaucratic character of the one and the city-state democracy of the other.[1] The Greeks had relatively little history of their own, they looked back with curiosity rather than reverence to the long ages of Babylonia and Egypt, the morality of which did not particularly apply to them; but they were very interested in proving a point by rigorous logical process *coram publico* in their assemblies, where every man was an equal citizen and every man could argue back. It was natural therefore for formal logic to develop among the philosophers of the Greek democracies. The Chinese philosophers were in a rather different position; they had indeed some important academies and societies for discussion among themselves,[2] but for the most part they frequented the courts of the reigning feudal princes as advisers and ministers. 'The Chinese philosopher,' wrote Stange, 'could not like his Greek counterpart discuss his ideas on a political situation with an assembly of men of equal rights on the same level as himself, he could only bring his thoughts to fruition in practice by gaining the ear of a prince. The democratic method of logical argumentation was not feasible in discussions with an absolute ruler, but an entirely different method, the citation of historical examples, could make a great impression. Thus it was that proof by historical examples prevailed very early in Chinese history over proof by logical argument.' There can be little doubt that from the social differences between the slave-owning city-state democracies of the ancient Western world on the one hand, and the feudal and proto-feudal bureaucratic states and empires of China on the other, far-reaching divergences in cultural development

[1] See his stimulating paper, H. O. H. Stange, 'Chinesische und Abendländische Philosophie; ihr Unterschied und seine geschichtlichen Ursachen', *Saeculum*, 1950, 1, 380.

[2] Notably the renowned Chi-Hsia Academy in the State of Chhi founded about 325 B.C., the group of scholar-scientists gathered together by Lü Pu-Wei between 260 and 240 B.C., and Liu An's group of natural philosophers active between 140 and 120 B.C. (cf. *SCC*, Vol. I, pp. 95 ff., 111, Vol. III, pp. 195 ff.).

will be explainable in this sort of way.[1] Essentially the Chinese method was analogical—like causes bring like effects, as it was then so it is now, and so it will be for ever. This faith was profound. Hence the great dominance of history (and all its ancillary sciences) throughout Chinese history. The idea that Europe was the only really history-minded civilization is in fact untenable—Clio was at least as much at home in Chinese dress.

MECHANICAL AND HYDRO-MECHANICAL TIME MEASUREMENT

In view of the appreciation of concrete time, celestial and terrestrial, in Chinese culture, it is perhaps not so remarkable, though the relation of China with chronometry has only recently been properly appreciated, that we owe to medieval artisans and scholars there the first solution of the problem of mechanical time-keeping. Six centuries of mechanical clockwork in China preceded the appearance of clocks in the European West.[2]

Sundials and clepsydras (dripping water clocks) were of course developed in ancient Babylonia and Egypt, spreading out from the Fertile Crescent in high antiquity all over the Old World. In

[1] One should not conclude from this that ancient Chinese society had no democratic elements—the case is quite contrary, and the effects were far-reaching, but it cannot be discussed here. Stange's emphasis on the role of city-state democracy in the development of formal logic in Greece recalls the similar proposal of Vernant to derive deductive geometry from the same social milieu. He regards the demonstration of geometrical propositions in the *agora* as a form of logical mathematics particularly congruent with an assembly of equal disputing participants, a democratical reasoning, just as the later Greek writing was a democratized script. Arithmetic and algebra were more specialized secretarial or bureaucratic techniques, less susceptible of public demonstration, and more like the old linear B script. Thus the preference for algebraic methods, so marked in China as well as Babylonia, would have gone naturally with the bureaucratic system of society, so different from the slave-owning city-state democracies. See J. P. Vernant, in *Scientific Change; Historical Studies in the Intellectual, Social and Technical Conditions for Scientific Discovery and Technical Invention, from Antiquity to the Present*, ed. A. C. Crombie (London, 1963), p. 102; and *Les Origines de la Pensée Grecque* (Paris, 1964).

[2] On the subject of this section see J. Needham, Wang Ling & D. J. de S. Price, *Heavenly Clockwork, the Great Astronomical Clocks of Mediaeval China* (Cambridge, 1960), as also *SCC*, Vol. IV; pt. 2, pp. 435 ff.

China the former were generally equatorial, never developing the complexity of Arabic and Western gnomonics because of the absence of Euclidean deductive geometry, but they gave rise to many a proverb similar to our own—'an inch of gold will not buy an inch of time (lit. light-and-dark)' (*tshun chin nan mai tshun kuang yin*), 'a foot of jade is no treasure, but one should struggle for an inch of shadow' (*chhih pi fei pao, tshun yin shih ching*).[1] The clepsydra evolved much further; adopting the inflow type the Chinese stabilized pressure-heads by multiplying the number of superimposed vessels and then by using constant-level overflow devices. They also had the receiver weighed continuously on a steelyard, and ended by weighing the intermediate vessel. This was probably the prelude to mounting a whole series of such vessels on a rotating wheel, and so the great break-through in accurate time-measurement came about.[2]

The invention of the mechanical clock was one of the most important turning-points in the history of science and technology, indeed of all human art and culture.[3] The problem was to find a way of slowing down the rotation of a set of wheels so that it would keep step with the great clock of the skies, that apparent diurnal rotation of the heavens which star-clerks and astronomers had studied since the beginning of civilization. The escapement was the first great achievement in the control of power. The mechanical clock, so familiar to all of us today, was truly at its birth a cardinal triumph of human ingenuity; it constituted perhaps the greatest tool of the Scientific Revolution of the seventeenth century, it trained the craftsmen who were needed for making the apparatus of modern experimental technique, and it furnished a philosophical model for the world picture which grew up on the basis of the 'analogy of mechanism'. But when exactly was it born? Until lately books on the history of time-keeping used to begin with a couple of chapters on sundials and

[1] On the sundial in China see *SCC*, Vol. III, pp. 302 ff.
[2] On the clepsydra in China see *SCC*, Vol. III, pp. 313 ff. together with Needham, Wang & Price, op. cit., pp. 85 ff.
[3] J. L. Synge, 'A Plea for Chronometry', *New Scientist*, 1959, **5**, 410, has well said that of all measurements made in physics that of time is the most fundamental.

clepsydras, then passing with a mortal leap to the invention of the verge-and-foliot escapement of the mechanical clocks of early fourteenth-century Europe. Some link was very obviously missing. And indeed we now know that an effective escapement was first invented at least 600 years earlier, at the eastern, not the western, end of the Old World. The fact that it was for a hydro-mechanical clock shows precisely the nature of the link.

This knowledge came forth from the recent study of a book written by one of the greatest Sung statesmen, also a naturalist and an astronomer, Su Sung (A.D. 1020 to 1101). In the year 1090 he took up his brush to give a monographic description of an elaborate astronomical clock-tower which during the previous two years had been erected under his supervision and with the collaboration of an engineer, Han Kung-Lien, at the capital, Khaifêng. It is entitled *Hsin I Hsiang Fa Yao* (New Design for an Armillary Clock). The first two chapters deal with the sphere and globe while the third describes the horological machinery in great detail. Since the observational instrument on the top storey was mechanized, as well as the globe on the first floor, and all the jack-work which manifested itself at each storey of a pagoda-like time-annunciator, this was the first astronomical clock-drive in history. Necessarily also it embodied the first solar-sidereal conversion gear. Power was derived not from a falling weight as in later Europe but from the torque of a water-wheel with scoops like a mill-wheel or Pelton turbine. The escapement which checked the forward motion of this wheel was a device of weighbridges and linkwork which remained stationary while each scoop was filling, but then operated instantaneously so as to open a gate and release one spoke, the next scoop being brought into position under the constant-flow water-jet. Steady motion was thus secured by intersecting the progress of a powered machine into intervals of equal duration—an invention of genius.

But it was not Su Sung's own. Once his technical terminology had been understood it was possible to trace back through earlier literature the records of the building of such hydro-mechanical clocks. A key point was found in the year A.D.

725 in the Thang dynasty, when the first of these escapements was devised, by a Tantric Buddhist monk I-Hsing, probably the greatest mathematician and astronomer of his age, and a military engineer Liang Ling-Tsan. The presence of the linkwork escapement in their clocks is certain, as also is that of luni-solar orrery gearing. The possibility still remains open that the escapement may go even further back, for from the time of Chang Hêng (A.D. 78 to 139) in the Han onwards, many texts tell of the automatic rotation of celestial globes in time with the heavens. They are not explicit, unfortunately, on the mechanisms of these 'proto-clocks'. It is not at all difficult to understand why these developments should have taken place in China so long ahead of Europe, for the expression of star positions on equatorial co-ordinates (equivalent to modern declination and right ascension) was standard practice there, not the ecliptic co-ordinates of the Greeks.[1] All stellar motion follows the former, while along the latter nothing moves, so that it was quite natural to wish to reproduce the motion in model form, 'for aid in computation'. What we still do not know is whether the Chinese escapement system was used in Europe during the century or so preceding the appearance of the verge-and-foliot falling-weight clocks, but there are some grounds for thinking so. At the least, the knowledge that the problem of mechanical timekeeping had been successfully solved elsewhere may have been inspiration enough for the first makers of mechanical clocks in Europe.[2]

Western writers have had much to say about the 'timeless Orient', but whatever other civilization their words may have applied to, it was not China. It is impossible to imagine that the vast works of erudition in Chinese literature, only a very few of which can be mentioned in this lecture, could have been brought

---

[1] Cf. Fig 7 on p. 78 above.

[2] Fig. 31 (p. 278) modified (with his agreement, and consultation with J. H. Combridge) from F. A. B. Ward, 'How Timekeeping became Accurate', *Chartered Mechanical Engineer*, 1961, **8**, 604, shows how the Chinese water-wheel linkwork escapement was a good deal more accurate than the early verge-and-foliot clocks of Europe. Its level was probably not surpassed until after the introduction of the pendulum about the middle of the seventeenth century and the further improvements that followed therefrom.

to completion unless their authors and team-leaders had had 'an eye on the clock'. A book ascribed to the great literary critic Liu Hsieh (d. *c.* A.D. 550), the *Hsin Lun* (*New Discourses*), has an interesting section entitled Hsi Shih (Sparing of Time).[1] Here we read:

'The worthies of old, wishing to spread abroad benevolence and righteousness in the world, were always struggling against time. They set no value on whole foot-lengths of jade, but a tenth of an inch of shadow (on the dial) was as precious as pearls to them. Thus it was that Yü the Great[2] raced with time to finish his work and paid no attention to the enquiries of Nanjung.[3] Thus it was that Tao Chung never stopped walking till the soles of his feet were as hard as iron. Confucius grudged every moment lost from reading, and Mo Ti was up and about again before his bed had had time to get warm. All these applied their virtue and genius to relieve the miseries of their times, so that they have left a good name behind them through a hundred generations.'

BIOLOGICAL CHANGE IN TIME

We have now said something of philosophy, history, chronology and horology. We must next enquire what went on in this endless chain of time which the Chinese took so seriously. First, what of the position of biological change and evolution? As soon as one looks at the ideas of traditional Chinese culture on living things, one finds that they never had any belief in the fixity of species. This followed because they never had any conception of special creation, and that was because creation *ex nihilo* by a Supreme Deity was itself unimagined by them; consequently there was no reason to believe that different kinds of living things could not turn into each other quite easily, given sufficient time. Careful observation would show what did or did not happen. Thus on one side the Chinese view of life was far more open than that of medieval or even eighteenth-century Europe, but on another side their Stoic-Epicurean non-creationism precluded them from a conception which in the West proved (at least in certain periods) favourable to the growth of the natural sciences, namely that of

[1] Reproduced in *Thu Shu Chi Chhêng*, Jen shih tien, ch. 4.
[2] Semi-legendary culture-hero and hydraulic engineer. Cf. p. 181 and Fig. 26.
[3] A character in *Chuang Tzu*.

Laws of Nature laid down by a supreme celestial lawgiver.[1] Without a more or less personal Creator one could not think of a divine Legislator for animals, plants and minerals as well as men; the operations of the Tao were in a way more mysterious, even though certain clear regularities (*chhang tao*) would certainly reveal themselves to 'faithful and magnificent' observers and experimenters, of whom there was no lack.

Thus the lore of metamorphoses was even more prominent in Chinese literature than in that of the West.[2] Numerous texts may be cited to show the acceptance of the possibility of slow evolutionary modifications and interconversions.[3] Recognition of the *scala naturae* developed among Warring States philosophers contemporary with Aristotle (fourth century B.C.), and there was an independent elaboration of the 'ladder of souls' theory with different and less animistic terminology, as in the *Hsün Tzu* book (third century B.C.). Recognition of the animal (and even vegetal) relationships of man led during the first millennium A.D. to a resolution of those controversies about human nature, which had so occupied the philosophers of the late Chou and Han periods, in terms of animal-like components within man. Thus Tai Chih about A.D. 1260 saw that the more highly social tendencies of man were peculiar to him, while his anti-social tendencies had to do with those elements of his nature which he shared with the lower animals.[4] Among the Neo-Confucians this led to a marked interest in comparative animal psychology, where 'gleams of righteousness' (*i i tien*) might perhaps be perceived.[5] A direct statement of

[1] On the general problem of the origin and development of the idea of Laws of Nature in the different Old World civilizations see *SCC*, Vol. II, pp. 518 ff. A slightly revised version of this account will be found in J. Needham, 'Human Law and the Laws of Nature', art. in *Technology, Science and Art; Common Ground* (Hatfield, 1961) repr. p. 299 below.

[2] No adequate treatment of the history of biology in Chinese culture either in Chinese or a Western language as yet exists, but we hope to present a balanced review of it in *SCC*, Vol. VI.

[3] For details of the instances referred to in this paragraph see J. Needham & D. Leslie, 'Ancient and Mediaeval Chinese Thought on Evolution', *Bull. Nat. Institute of Sciences of India*, 1952, **7** (Symposium on Organic Evolution), 1.

[4] Cf. *SCC*, Vol. II, pp. 21 ff.

[5] *SCC*, Vol. II, pp. 488 ff.

evolutionary transformation is found in a famous passage of the *Chuang Tzu* book (fourth century B.C.), though several of the species mentioned there are not now identifiable. In this book too we have a view of biological changes arising as adaptation to particular environments, and an adumbration of the idea of natural selection in passages which point out the 'advantages of being useless'.[1] Copious biological discussions occur in the book by the great sceptic Wang Chhung entitled *Lun Hêng* (Discourses Weighed in the Balance) and written about A.D. 83. He insists that man is an animal like other animals, though the noblest of them, rejects mythological birth stories but not spontaneous generation, maintains that all transformations, however weird, are fundamentally natural, and speaks of 'sports', genetic inheritance, animal migrations and tropisms. After the spread of Buddhism in China, interest in the philosophy of metempsychosis led to renewed study of embryological and metamorphic time-processes. Early in the twelfth century A.D. Chêng Ching-Wang tried to analyse certain believed natural transformations, linking them with the ladder of souls and interpreting them in the light of Buddhist migrations between the lower and higher levels of being. Good actions authorized some spirits to rise in the scale, while soteriological virtue (as well as evil actions) impelled others to descend.[2]

Evolutionary naturalism came fully into focus in the thought of the great Neo-Confucian school, a movement of systematization quite close in date to that of the scholastic philosophers of Europe, with whom these Chinese thinkers are often compared. Just as the Europeans sought to harmonize Greek philosophy with the doctrines of Christian theology, so the Neo-Confucians drew upon all the older philosophies, Confucianism, Taoism and Buddhism for their own synthesis. But their spirit of organic materialism was so different from that of the European scholastics that Chu Hsi (A.D. 1131 to 1200), their greatest figure, has been termed with at least equal enthusiasm the Herbert Spencer as well as the Thomas Aquinas of China. To understand the universe, as man sees it, the

[1] I.e. to one's predators. *SCC*, Vol. II, pp. 78 ff.
[2] *SCC*, Vol. II, pp. 421 ff.

Neo-Confucians worked with two fundamental concepts only, *chhi*, or what we should now call matter-energy, and *li*, the principle of organization and pattern in all its forms. It was extraordinary that they could reach this economy of principle, this world-view so congruent with modern science, in a civilization which not only had not developed modern science, but was destined not to be able spontaneously to develop it. For the Neo-Confucians the universe was essentially moral, not because there existed beyond space and time a moral personal deity controlling his creation, but because the universe was one which had the property of bringing to birth moral values and ethical behaviour when that level of organization had been reached at which it was possible that they should manifest themselves. Organization in the animal kingdom begins to approach this, very incompletely and one-sidedly (hence the gleams of righteousness), but it is only with the fully developed nervous system of gregarious social man that the universe manifests ethical values. Thus long before the Darwinian age evolutionary naturalism was very clearly stated by Chinese philosophers. But they envisaged a whole succession of these phylogenetic unfoldings rather than one single evolutionary series.

This was doubtless a legacy from Indian thought mediated through Buddhism, envisaging successive time-spans, finite but enormously long, which included the *kalpa* and the *mahākalpa*.[1] All the Neo-Confucians accepted the idea that the universe passed through alternating cycles of construction and dissolution. It seems to have been first systematized by a Taoist precursor of Neo-Confucianism, Shao Yung (A.D. 1011 to 1077) who applied the duodenary series of cyclical characters (cf. p. 232, above) to its various phases.[2] Chu Hsi was probably led to his remarkably correct views on the nature of fossils, and other Sung scholars such as Shen Kua (A.D. 1031 to 1095) to their penetrating insights into

[1] See Zimmer, H. *Philosophies of India*, pp. 224, 226, on the Jaina cycles; and also his *Myths and Symbols in Indian Art and Civilisation* ed. J. Campbell (New York, 1946), pp. 11 ff., 16 ff., 19 ff. Here he discusses the *kali-yuga*, last and worst of the four world ages. Cf. p. 288 below.

[2] Cf. *SCC*, Vol. II, p. 485, also Vol. IV, pt. 1, p. 11.

mountain-building and erosion, foreshadowing the 'plutonic' and 'neptunian' ideas of the early nineteenth century, by their meditations on the recurrent world-catastrophes or cataclysms in which they believed.[1] Other thinkers, such as Hsü Lu-Chai (A.D. 1209 to 1281) applied the hexagrams of the *Book of Changes* to the phases of the evolutionary cycle, and Wu Lin-Chhuan (A.D. 1249 to 1333) estimated its length in time as 129,600 years.[2] Like the Thang calculations of astronomical periods in millions of years,[3] these world-views were immensely more spacious than those of seventeenth- and eighteenth-century Europe, with its fixing of the date of creation at 22 October, 4004 B.C. at six o'clock in the evening.[4] Biological and social evolution were thus conceived in a cyclical setting, and would for ever continue to recur, each cycle being separated by a kind of Ragnarök, a twilight of the gods, the reduction of everything to a disordered and chaotic state, after which all things slowly evolved anew. One might say that the single action of the world's drama, as we think of it today, was replaced by a whole series of repeat performances. And while the rise was slow, the downfall was rapid. The Neo-Confucians would have appreciated the words of William Harvey on individual beings:

'For more, and abler, operations are required for the fabrick and erection of living beings, than for their dissolution and the plucking of them down, for those things that easily and nimbly perish, are slow and difficult in their rise and complement.'

---

[1] See *SCC*, Vol. III, pp. 598 ff., 603 ff.

[2] *SCC*, Vol. II, pp. 486 ff., Vol. III, p. 406.

[3] See *SCC*, Vol. II, p. 420, Vol. III, pp. 120, 408. In A.D. 724 the great monk-astronomer I-Hsing computed the number of years which had then elapsed since the 'Grand Origin' (Thai Chi Shang Yuan) or general conjunction of the planets, and obtained a result of 96,961,740 years. The fact that a general conjunction is impossible is irrelevant to the spaciousness of the time-periods which the Chinese medieval astronomers were prepared to envisage.

[4] This was the celebrated reckoning of the learned James Usher, Archbishop of Armagh. See his *Chronologia Sacra* (Oxford, 1660), p. 45, and *Annals of the World* (Oxford, 1658), p. 1. I am indebted to Prof. H. Trevor-Roper for assisting me with these references.

TIME AND SOCIAL DEVOLUTION OR EVOLUTION, TA THUNG AND THAI PHING

We have been speaking of social evolution, but it was implicit in the Neo-Confucian world-view rather than clearly defined. Chinese thinkers were rather divided on the question of what had happened to human society in time, and there were two sharply contrasting attitudes. On the one hand there was the conception of a Golden Age of primitive communalism or of sage-kings from which mankind had steadily declined,[1] while on the other there was a recognition of culture-heroes as progenitors of something much greater than themselves, with an emphasis on development and evolution out of primitive savagery.[2]

The first of these views was characteristic of the ancient Taoist philosophers, and in them it was closely connected with a general opposition to proto-feudal and feudal society.[3] They harked back always to the ancient paradise of generalized tribal nobility, of co-operative primitivity, of spontaneous collectivism ('When Adam delved and Eve span, Who was then the gentleman?'), before the aenolithic differentiation of lords, priests, warriors and serfs. They were probably stimulated in this by the persistence of pre-feudal relationships among some of the tribal peoples on the fringe of Chinese society, such as those who have been known as the Miao, Chiang, Lo-lo and Chia-jung in our own time, and are only now, after more than two millennia, being integrated into Chinese society as a whole. And indeed many traces of their basic

[1] For example the great second-century B.C. medical classic *Huang Ti Nei Ching Su Wên* (ch. 14) periodized history into ancient (*shang ku*), middle-old (*chung ku*) and recent (*tang chin*) ages, saying that there had been a gradual decline in men's resistance to diseases, so that stronger drugs and treatments were required as time went on.

[2] One entire chapter of the *Huai Nan Tzu* book (*c.* 120 B.C.) is devoted to proving social change and progress since the most ancient times, with many references to material improvement (ch. 13, Morgan tr., pp. 143 ff.). The *Huai Nan Tzu* is very Taoist in many ways, but this was a viewpoint of Han Taoism rather than of that of the Warring States.

[3] Cf. *SCC*, Vol. II, pp. 86 ff., 99 ff., 104 ff., 115 ff.

fourth-century B.C. opposition to feudal and feudal-bureaucratic society continued to cling to the Taoists all through Chinese history, long after their school had generated a mystical nihilism for the educated scholars and an ecclesiastical organization for the poor peasants. Their continued presence in the background of agrarian rebellions under every dynasty is alone evidence of this; they were in a perpetual opposition which only the equalitarian socialism of our own time would satisfy.[1] Of course there are many European parallels for the Taoist idea of a Golden Age—the Cronia and the Saturnalia of Rome commemorating the vanished ages of Cronos and Saturn, the repudiation of over-civilized life by the Stoics and Epicureans, the Christian doctrine of the Fall of man (perhaps deriving from the ancient Sumerian laments for lost social happiness in lordless society), the stories of the 'Isles of the Blest', and finally the eighteenth-century admiration for the Noble Savage, stimulated by the first contacts of Westerners with the real-life 'paradises' of the Pacific.[2]

By some literary accident the most famous statements of the Taoist theory of regressive devolution occur in books of other schools, the second-century B.C. *Huai Nan Tzu*, and the first-century B.C. Confucian *Li Chi* (Record of Rites). Here we shall quote a passage from the Li Yün chapter of the latter.[3]

'When the Great Tao prevailed, the whole world was one Community (*thien hsia wei kung*).[4] Men of talents and virtue were chosen (to lead the people); their

[1] Cf *SCC*, Vol II, p. 60.

[2] Cf. *SCC*, Vol. II, pp. 127 ff.

[3] Ch. 9, Legge tr., vol. 1, pp. 364 ff., here modified from *SCC*, Vol. II, p. 167. Li Yün may be translated 'The Mutations of Social Institutions'. The wording of parallel passages in the *Mo Tzu* book, chs. 11, 12, 13, 14, 15 (tr. Mei Yi-Pao, pp. 55, 59, 71, 80, 82), fix the date as fourth century B.C., not first century A.D. But these passages are 'progressive' rather than 'regressive' in tendency, criticizing the ancient ruler-less times as an age when humanity was all 'at sixes and sevens', and placing the Ta Thung state in the future, to be brought about by the practice of universal love (*chien ai*). The actual expression Ta Thung is not used in *Mo Tzu*. A similar account to that in the *Li Chi*, but much shorter, occurs in the *Huai Nan Tzu* book (120 B.C.), ch. 2, where the expression Ta Chih (the Ideal Rule) is used instead of Ta Thung; cf. Morgan tr. p. 35.

[4] Lit. 'for the general use', i.e. not the property of the emperor, feudal lords and patrician families.

words were sincere and they cultivated harmony. Men treated the parents of others as their own, and cherished the children of others as their own. Competent provision was made for the aged until their death, work was provided for the able-bodied, and education for the young. Kindness and compassion were shown to widows, orphans, childless men and those disabled by disease, so that all were looked after. Each man had his allotted work, and every woman a home to go to. They disliked to throw valuable things away, but that did not mean that they treasured them up in private storehouses. They liked to exert their strength in labour, but that did not mean that they worked for private advantage. In this way selfish schemings were repressed and found no way to arise. Thieves, robbers and traitors did not show themselves, so the outer doors of the houses remained open and were never shut. This was the period of the Great Togetherness (Ta Thung).[1]

But now the Great Tao is disused and eclipsed. The world (the empire) has become a family inheritance. Men love only their own parents and their own children. Valuable things and labour are used only for private advantage. Powerful men, imagining that inheritance of estates has always been the rule, fortify the walls of towns and villages, and strengthen them with ditches and moats. "Rites" and "righteousness" are the threads upon which they hang the relations between ruler and minister, father and son, elder and younger brother, and husband and wife. In accordance with them they regulate consumption, distribute land and dwellings, raise up men of war and "knowledge"; achieving all for their own advantage. Thus selfish schemings are constantly arising, and recourse is had to arms; thus it was that the Six Lords (Yü "the Great", Thang, Wên, Wu, Chhêng and the Duke of Chou) obtained their distinction. . . . This is the period which is called the Lesser Tranquility (Hsiao Khang).'

The Mohists undoubtedly sympathized to some extent with this account of the ideal co-operative, even socialist, society, which had, it was thought, existed in the remote past, but it was certainly not part of Confucian ideology at all. Nevertheless in spite of the later universal dominance of Confucianism in Chinese life, the idea of

[1] This phrase, which we might equally well translate the Great Community, was also used in a rather different sense by the late Warring States philosophers, namely to indicate the parallel of the Microcosm (man) with the Macrocosm (the universe). For an example of this see Lü Shih Chhun Chhiu (239 B.C.), ch. 62 (R. Wilhelm tr., p. 160). But the senses are not so far apart because the ancient Chinese felt that social community was 'intended by Nature' and that class differentiation and all strife was a violation of the natural order, a violation moreover, which would upset nature and lead to natural calamities, or at least to unfavourable weather conditions, epidemics, etc. Besides, thung as 'with-ness' allows the translation 'Great Similarity'.

the Ta Thung society enjoyed a certain immortality, for if it had really once existed upon the face of the earth, it might perhaps be brought into existence again.[1] Indeed, Confucianism itself, with its emphasis upon development and social evolution, contributed to this very end. And although the innumerable peasant rebellions through Chinese history rarely pushed their thinking beyond the establishment of a new and better dynasty,[2] at the same time their more visionary elements were often inclined to reverse the time-dimension of the regressive conception and turn it into a progressive one. Nineteen centuries later than the Han, in our own time, these two little words had vastly gained, not lost, in numinous, emotional and revolutionary force.[3]

There was indeed a parallel (or rather, inverse) ascending Confucian sequence, but before examining it we must take a look at another, related, conception, that of Thai Phing (the Great Peace and Equality).[4] This was another 'phrase of power', but widely varying in interpretation.[5] The Golden Age and the realizable Utopia are here not very clearly dissociable; it is hard to find definite statements in ancient texts that this was an era only in the far past which could never come back, or that it was purely something to look forward to in the future. Undoubtedly many imperial reigns were consciously trying to attain it. The term appears first in 239 B.C. in the *Lü Shih Chhun Chhiu* (Master Lü's Spring and Autumn Annals), a famous compendium of natural philosophy, where it denotes a state of peace and prosperity which

[1] Cf. the valuable paper of Hou Wai-Lu, 'Socialnye Utopii Drevnego i Srednevekovogo Kitaia (Social Utopias of Ancient and Mediaeval China)', *Voprosy Filozofii*, 1959, **9**, 75.

[2] See Shih Yu-Chung, 'Some Chinese Rebel Ideologies', *Thoung Pao*, 1956, **44**, 150.

[3] On the history of the Ta Thung concept in China there is a valuable little book by Hou Wai-Lu, Chang Kai-Chih, Yang Chao & Li Hsüeh-Chin, *Chung-Kuo Li-Tai 'Ta Thung' Li Hsiang* (Peking, 1959).

[4] The word *phing* has both meanings.

[5] It is now under intensive study by sinologists, historians and social philosophers. The best review of the subject in a Western language is probably that of W. Eichhorn, 'Thai-Phing und Thai-Phing Religion', *Mitt. d. Inst. f. Orientforschung*, 1957, **5**, 113, seconded by T. Pokora, 'On the Origins of the Notions of Thai-Phing and Ta-Thung in Chinese Philosophy', *Archiv. Orientalní*, 1961, **29**, 448.

can be brought about magically by music in harmony with the cyclical operations of Nature.[1] During the following centuries the emphasis was sometimes upon social peace springing from the harmonious collaboration of different social classes each contented with its lot, sometimes upon a harmony of natural phenomena (which man could perhaps induce) leading to an abundance of the kindly fruits of the earth, and sometimes upon the idea of equality, with undertones of reference to that primitive classless society which might in the last day be restored. Some thought that the Great Peace had existed under the sage-kings of high antiquity, others that it was attainable by good imperial government here and now, and others again that it would come to pass at some future time. It is worth while to cite some of these different opinions.

The mysterious social magic of Master Lü appears again in the chapter on rites in the *Chhien Han Shu* (History of the Former Han Dynasty), *c.* A.D. 100, where it is said that the full application of the rites of the former kings will bring about the Thai Phing state.[2] This had particular reference to the seasonal ceremonies of the Ming Thang, or cosmic temple, where the emperor and his assistants carried out liturgical observances before Heaven on the people's behalf. The biography of the minister Tou Ying (d. 131 B.C.) tells of his support for the Ming Thang and other ceremonial measures as the way in which the Great Peace could be attained,[3] and the *Huai Nan Tzu* book (120 B.C.) specifically connects its attainment with the ritual purity and clarity of the cosmic temple services.[4] On the other hand the biography of Tungfang Shuo (d. *c.* 80 B.C.) speaks of the induction of natural conditions favourable to mankind by man's own social harmony,[5] and the economic chapter of the *Chhien Han Shu* goes so far as to apply the term Thai Phing to record-harvest years.[6] One of the sections of the *Chuang*

---

[1] Ch.22, R. Wilhelm tr., p. 56.
  Ch.22, W. Eichhorn, loc. cit., p. 116.
[3] *Chhien Han Shu*, ch. 52, W. Eichhorn, loc. cit., p. 123.
[4] Ch. 2, W. Eichhorn, loc. cit., p. 123.
[5] *Chhien Han Shu*, ch. 65.
[6] Ch. 24.

Tzu book which is probably a Han interpolation says that the highest aim, good government, Thai Phing, is to be attained not by human skill and planning, but only by following the Tao of Heaven. Now 'the Tao of Heaven is to revolve ceaselessly and not to amass virtue or things in any particular place, thus it is that all things are brought into perfection by it (*thien tao yün erh wu so chi, ku wan wu chhêng*).'[1] Here at once is the theme of Great Equality as well as Great Peace, and echoes awake throughout the Taoist writings. The 'equality of things and opinions' was the doctrine of the Chi-Hsia Academicians Phêng Meng, Thien Phien, and Shen Tao (all *fl.* 320 to 300 B.C.),[2] as well as the title of a genuine chapter of the *Chuang Tzu* book (*c.* 290 B.C.) which contains some of Chuang Chou's clearest keys to Taoist epistemology, scientific world-outlook and democratic social thinking.[3] 'The great highway (of the Tao of justice and righteousness)', says the *Tao Tê Ching* 'is broad and level (*ta tao shen i*)'—one of those pregnant sayings which recalls Hebrew prophecy, 'Make straight the way of the Lord', and touches the mystical poetry of road engineering in all ages and peoples, 'the valleys shall be exalted, and the mountains shall be made low.'[4] Its equalitarian meaning cannot be in doubt for the poem goes on to castigate the feudal lords for amassing wealth and oppressing the peasant-farmers; 'these are the riotous ways of brigandage, these are not the great highway.'[5]

Many ancient texts, however, speak of Thai Phing only as the Golden Age of the sage-kings of high antiquity; so Chia I in his *Hsin Shu* (*c.* 170 B.C.),[6] the alchemist Wu Pei talking with his

[1] Ch. 13 (Thien Tao), Legge tr., vol. I, pp. 330, 337.
[2] See Fêng Yu-Lan, *A History of Chinese Philosophy*, vol. I, pp. 153 ff. These men were all members of the Chi-Hsia Academy founded by Prince Huan of Chhi, (*c.* 325 B.C.).
[3] Ch. 2 (Chi Wu Lun), Legge tr., vol. I, pp. 176 ff.
[4] Cf. Isaiah, XL, 3, 4. There was also a mystique of topographic levelling in Hindu and Buddhist thought, the 'alluvial' flatness left behind by world floods or catastrophes (cf. p. 263 below) on which the Buddhas and Bodhisattvas pace. On this idea see P. Mus, 'La Notion de Temps Réversible dans la Mythologie Bouddhique', *Annuaire de l'Ecole Pratique des Hautes Etudes (Sect. des Sciences Religieuses)*, 1939, pp. 15, 33 ff., 36.
[5] Ch. 53, Wu tr. p. 75, Chhu tr. p. 66, Duyvendak tr. p. 117.
[6] Ch. 52 (Hsiu Chêng Yü), tr. Eichhorn, loc. cit., p. 118.

patron the Prince of Huai-Nan (c. 130 B.C.),[1] the chapter on rites in the
Shih Chi (c. 100 B.C.),[2] and the Yin Wên Tzu book.[3] Others make
it clear that in certain prosperous periods the Great Peace was considered as already having been attained. It is clear that the first
emperor Chhin Shih Huang Ti was explicitly striving for it,[4] and
by 210 B.C. claimed to have inaugurated it; an inscription set up
in that year says: 'The people are pleased with the standard rules
and measures, and felicitate each other on the preservation of the
Great Peace.'[5] So also Lu Wên-Shu (fl. 70 B.C.) considered that
the reign of Han Wên Ti (179 to 157 B.C.) had been a period of
Thai Phing.[6] All these different opinions can be found discussed in
the Lun Hêng (A.D. 83); one chapter records its attribution to the
time of the ancient sages Yao and Shun, another says that many
believed it was presaged by the appearance of the phoenix and the
unicorn, and in a third Wang Chhung gives his own belief that
Thai Phing prosperity had occurred several times during the two
Han dynasties.[7]

We come now to the incorporation of the Thai Phing concept
into a temporal sequence analogous to that of the Ta Thung. It
arose out of the exegesis of the Chhun Chhiu (Spring and Autumn
Annals) by the scholars of the Han. This book was a chronicle of
the feudal State of Lu between 722 and 481 B.C., and there was a

[1] Chhien Han Shu, ch. 45.
[2] Ch. 23; Chavannes tr., vol. 3, p. 211.
[3] Ch. 2 (Ta Tao), where Thien Phien, lecturing on the Shu Ching (Historical
Classic), said that in the time of the (legendary) emperor Yao, there had been Thai
Phing.
[4] Shih Chi, ch. 6; Chavannes tr. vol. 2, p. 180.
[5] Shih Chi, ch. 6; Chavannes tr. vol. 2, p. 189.
[6] Chhien Han Shu, ch. 51.
[7] Respectively, ch. 26 (Ju Tsêng), Forke tr., vol. 1, p. 494; ch. 50 (Chiang Jui),
Forke tr., vol. 1, p. 364; and ch. 57 (Hsüan Han), Forke tr., vol. 2, pp. 192 ff. Wang
Chhung also combated the excessive veneration of the sages and the belief in a
Golden Age in his ch. 56 (Chhi Shih—that all generations are much the same),
Forke tr., vol. 1, pp. 471 ff. It is interesting that the phrase Thai Phing occurs in
quite a number of place-names, and even more that it was used as the appellation
of no less than six reign-periods. These were in the following dynasties, San Kuo
(Wu) A.D. 256, Northern Yen A.D. 409, Northern Wei A.D. 440, Liang A.D. 556,
Sung A.D. 976, and Liao A.D. 1021. Ta Thung was also used in reign-period names,
twice, Liang, A.D. 535 to 546, and Liao, A.D. 947.

persistent tradition that Confucius himself had edited it. It has come down through the ages accompanied by commentaries in three traditions known as the *Tso Chuan*, the *Kuliang Chuan* and the *Kungyang Chuan*.[1] Master Tsochhiu's Enlargement carried the history down a little further, to 453 B.C., and was compiled from ancient written and oral traditions of several States (not only Lu) between 430 and 250 B.C., though with many later changes and additions by Confucian scholars of the Chhin and Han. Master Kuliang's Commentary and Master Kungyang's Commentary differed from this in that they were not formed partly from independent ancient historical writings, but restricted themselves to word-for-word explanations of the chronicle text.[2] The importance of this was the belief, already mentioned (p. 240), that great moral weight attached to the precise terms which Confucius had used in each given historical circumstance. During the second and first centuries B.C. the scholars of the Han formed groups which specialized in the study of one or other of these traditions, indeed separate chairs were established for them in the imperial university.[3] Among those learned in the tradition of Master Kungyang was that remarkable philosopher (who made his mark in many other ways), Tung Chung-Shu (179 to 104 B.C.). Tung developed a theory of the San Shih or Three Ages, grouping the events in the *Chhun Chhiu* into a triple classification, those that Confucius himself had personally witnessed (541 to 480 B.C.), those that he

[1] A good introductory account of this literature will be found in P. van der Loon, 'The Ancient Chinese Chronicles and the Growth of Historical Ideals', art. in *Historians of China and Japan*, ed. W. G. Beasley & E. G. Pulleyblank (London, 1961), p. 24. For fuller details the monograph of Wu Khang mentioned on p. 265 below, may be studied.

[2] All three were supposed to derive from the oral teaching of Confucius himself.

[3] This dates from 124 B.C. though the governmental title of Po-Shih (doctor or Professor) had appeared already in the third century B.C. and the principle of imperial examinations in 165 B.C. When Han Wu Ti endowed 'disciples' (*ti-tzu*) as well as professors, the imperial university may be said to have been established. By 10 B.C. it had as many as 3,000 students, not all, of course, 'on the foundation'. The term Thai Hsüeh, afterwards borne for centuries by the university, occurs first in a memorial by Tung Chung-Shu himself urging its establishment, though the emperor preferred the plans of Kungsun Hung for the same design.

heard of from oral testimony (626 to 542 B.C.), and those that he knew only through written records (722 to 627 B.C.).[1] Then in the Later Han this was converted into an ascending social evolutionary series, first applied to the Confucian redaction, and afterwards extended to a universal application. Here the key mind was Ho Hsiu (A.D. 129 to 182) whose work became the standard commentary on the *Kungyang Chuan*.[2] He wrote:

'In the age of which he heard through transmitted records, Confucius saw (and made evident) that there was an order arising from Weakness and Disorder (Shuai Luan),[3] and so directed his mind primarily towards the general (scheme of things). He therefore considered his own State (of Lu) as the centre, and treated the rest of the Chinese oikoumene as something outside (his scheme). He gave detailed treatment to what was close at hand, and only then paid attention to what was further away....

In the age of which he heard through oral testimony he saw (and made evident) that there was an order arising of Approaching Peace (Shêng Phing). He therefore considered the Chinese oikoumene as the centre, and treated the peripheral barbarian tribes as something outside (his scheme). Thus he recorded even those assemblies outside (his own State) which failed to reach agreement, and mentioned the great officials even of small States....

Coming to the age which he (personally) witnessed, he made evident that there was an order (arising) of Great Peace (Thai Phing). At this time the barbarian tribes became part of the feudal hierarchy, and the whole (known) world, far and near, large and small, was like one. Hence he directed his mind still more profoundly to making a detailed record (of the events of the age), and therefore exalted (acts of) love and righteousness....'

Here then we have the formal simulacrum of a process of social

[1] See Fêng Yu-Lan, *History of Chinese Philosophy*, vol. 2, p. 81.

[2] Fêng Yu-Lan, op. cit., vol. 2, p. 83. The passage occurs at the end of the first chapter of the *Kungyang Chuan*.

[3] This phrase has an undertone of the Golden Age theory because *shuai* means decay or decadence as well as weakness and feebleness. But it is doubtful whether this was intended because an alternative form of the phrase found in many texts is Chü Luan, *chü* meaning forcible occupation or possession, the seizing of lands and goods, rebellion, etc., i.e. the correlate of weakness; in other words the state of society described in the Gospel of St Luke, 11.21, or the 'law of the fishes' in Buddhism, unending internecine strife, 'Nature red in tooth and claw'. Cf. Hsiao Kung-Chhüan, 'Khang Yu-Wei and Confucianism', *Monumenta Serica*, 1959, 18, 96, p. 142.

evolution in time, ready to be taken over into the general thought of the people as applicable to the whole of civilization.

Already before the time of Ho Hsiu, religious Taoism fermenting among the people had adopted the idea of Thai Phing in this way.[1] Much study is now being given to a corpus of ancient documents of which the chief is a book entitled *Thai Phing Ching* (Canon of the Great Peace), difficult to date because probably written at different times between the Warring States period (*c.* fourth century B.C.) and the end of the Later Han (A.D. 220).[2] Though the greater part of this is concerned with religious and superstitious practices, revelations and prophetic warnings, there are passages which link up with the revolutionary Taoism of the great national uprisings—the 'Red Eyebrows' of A.D. 24 led by Fan Chhung, and the 'Yellow Turbans' (A.D. 184 to 205) under Chang Chio.[3] It must of course be understood that the *Thai Phing Ching* and its associated texts were greatly expurgated in subsequent times by Taoists loyal to the established order.[4] But the popular religious Taoism of the Han was millenniarist and apocalyptic; the Great Peace was clearly in the future as well as the remote past. In the 'Canon' we hear of rural social solidarity, sins committed against the community and their forgiveness, an anti-technology complex, the overcoming of village feuds, and the particularly high place accorded to women. We also find a theory of cycles opposite in character to those of the Neo-Confucians already mentioned (p. 252),

[1] If Pokora, loc. cit., is right, the ideas of Ho Hsiu were directly derived from the popular progressive apocalyptic, possibly through the intermediation of Yü Chi (*c.* A.D. 120 to 200). Yü Chi was a naturalist, physician and thaumaturgist, one of the fathers of the Taoist church, and probably the author of one or more of the books which formed the material of the *Thai Phing Ching*.

[2] The corpus has been newly edited with the title *Thai Phing Ching Ho Chiao* (Peking, 1960) by Wang Ming, who attempts to reconstitute the original text of the main work. Pokora, loc. cit., gives a brief description of his book, and Eichhorn, loc. cit., discusses the contents of some of the documents which it includes.

[3] On him see W. Eichhorn, 'Bemerkungen zum Aufstand des Chang Chio und zum Staate des Chang Lu', *Mitt.d.Inst. f.Orientforschung*, 1955, **3**, 291.

[4] Nevertheless it still contains eloquent passages quite in the vein of that revolutionary thinker Pao Ching-Yen who (if not a literary creation of Ko Hung's) must have flourished in the latter part of the third century A.D.; see *SCC*, Vol. II, pp. 434 ff.

which throws light upon the practices of another great Taoist rebel leader Sun Ên (d. A.D. 402).[1] As the sins of mankind's evil generations increase to a climax, world catastrophes, flood and pestilence sweep all away—or nearly all, for a 'holy remnant' (a 'seed people', *chung min*), saved by their Taoism, win through to find a new heaven and a new earth of great peace and equality, under the leadership of the Prince of Peace (Ta Thai-Phing Chün), of course Lao Tzu. Then everything slowly worsens again until another salvation is necessary. Thus unlike the cycles of the Neo-Confucians which rose extremely slowly and ended in a flash, those of the religious Taoists issued fresh from chaos 'wie herrlich als am ersten Tag' and then fell slowly till the day of doom. But whether or not time was thought of as boxed this way in cyclical periods, the Thai Phing ideal was now for ever inscribed upon the banners of the Chinese people in one rebellion after another. It was clearly stated to be the aim of the Ming revolutionary Chhen Chien-Hu (c. A.D. 1425); and gave of course the name to the great Thai-Phing Thien-Kuo movement which between 1851 and 1864 nearly toppled the Manchu (Chhing) dynasty, and which is regarded in China today as the closest forerunner of the People's Republic.[2]

---

[1] On him see W. Eichhorn, 'Description of the Rebellion of Sun Ên and earlier Taoist Rebellions', *Mitt.d.Inst.f.Orientforschung*, 1954, **2**, 325, with an appendix, 'Nachträgliche Bemerkungen zum Aufstände des Sun Ên', p. 463.

[2] The standard modern work on this great but ultimately abortive revolution is by Lo Erh-Kang, *Thai-Phing Thien-Kuo Ko-Ming Chan Chêng Shih* (A History of the Revolutionary War of the Heaven-Ordained Kingdom of Great Peace and Equality), Peking, 1949; and since then eight volumes of source material have been edited by Hsiang Ta *et al.*, *Thai-Phing Thien-Kuo*, Peking, 1957. There is no satisfactory book on the subject as yet in any Western language, but three contemporary classics may be mentioned, the first by a British government interpreter, the second by a missionary, and the third by a soldier of fortune who fought with the Thai-Phing armies. T. T. Meadows, *The Chinese and their Rebellions, viewed in connection with their National Philosophy, Ethics, Legislation and Administration, to which is added, an Essay on Civilization and its Present State in the East and West* (Bombay and London, 1856, Stanford, Calif., n.d. (1953)), is a work in which the discursive background material (not in itself uninteresting) equals in amount the valuable first-hand description. W. H. Medhurst, *Pamphlets issued by the Chinese Insurgents at Nanking; to which is added a History of the Kwang-se [Kuangsi] Rebellion, gathered from Public Documents; and a Sketch of the Connection between Foreign Missionaries and the Chinese*

But this was not at all the end. One of the greatest reformers and representatives of modern Chinese thought, Khang Yu-Wei, who lived (1858 to 1927) throughout the period of intellectual strain when China was absorbing and digesting the new ideas which contact with the modern scientific civilization of the West had brought, drew greatly upon these age-old dreams and theories of progress. His classical studies led him to adopt positions which modern historical philology cannot now sustain,[1] but his

Insurrection; concluding with a critical view of several of the above Pamphlets (Shanghai, 1853), is naturally mainly concerned with the quasi-Christianity of the Thai-Phing revolutionaries. Lin-Le (Ling-Li, i.e. A. F. Lindley), Ti-Ping Tien-Kwoh; the History of the Ti-Ping Revolution, including a Narrative of the Author's Personal Adventures' (London, 1866), has more about the adventures than the history, but since the work of one who did something to counterbalance the military intervention of other foreigners (Ward, Burgevine and Gordon) on the imperialist side, it gives insights into the character of the Thai-Phing leaders otherwise unobtainable. Amongst recent publications there is the book of Chêng Chê-Hsi (J. C. Chêng), Chinese Sources for the Thai-Phing Rebellion, 1850 to 1864 (Hong Kong, 1963), but it is so lacking in commentary and explanation that it can serve only as a companion volume to other accounts. To this may be added a useful paper by Shih Yu-Chung (V. Y. C. Shih), 'The Ideology of the Thai-Phing Rebellion', Sinologica, 1951, 3, 1; and (especially interesting in view of what is said in the present article on the Judaeo-Christian attitude to time), E. P. Boardman's Christian Influence on the Ideology of the Thai-Phing Rebellion (Madison, Wis., 1952). Lastly see also G. Taylor, 'The Thai-Phing Rebellion, its Economic Background and Social Theory', Chinese Soc. & Polit. Sci. Rév., 1933, 16, 545.

[1] This is a complex question which can only be touched upon here. It involves the 'Old Text' and 'New Text' controversy which divided the scholars of the Han, and no less those of the late Chhing who delved again into Han studies. This division had arisen because of the discovery, during the second century B.C., of a set of versions of the classics (the Shu Ching or 'Historical Classic', the Shih Ching or Book of Odes, the Tso Chuan and the Chou Li) which differed from the texts previously accepted, and which were written in the archaic script of the early (Western) Chou. Traditionally this occurred during the destruction of the supposed house of Confucius in 135 B.C. when Prince Kung of Lu (Lu Kung Wang), Liu Yü, was enlarging his palace; but similar texts were also said to have been among those collected by the great bibliophile Liu Tê (d. 130 B.C.), the Prince of Ho-Chien (Ho Chien Wang). The terminology is rather confusing because the 'Old Texts' were those newly discovered in the Former Han, while the 'New Texts' were those which had the old authority of continuous use; one has to remember to think of them as the 'Old Script Texts' and the 'New Script Texts'. Many subsequent centuries of scholarly debate have ended in the conclusion that the story of the single discovery was a legend, and that some at least of the 'old versions' were probably forgeries, though the Shu Ching ones were not identical with the present 'Old

thought was deeply influenced both by the seemingly Mohist Great Togetherness (Ta Thung) and the Taoist Great Peace and Equality (Thai Phing). Interpreting them both in the ascending

---

Text' chapters which are known to have been compiled with ancient fragments about A.D. 320. From the point of view of the history of scientific thought the controversy has particular interest in that while the members of the New Text school (i.e. those who accepted the texts which had been continuously transmitted through the official teaching tradition) were textually on stronger ground, they accepted all the superstitious pseudo-sciences of the time, and with them that empirical open-mindedness in which the sprouts of experimental science could burgeon; and while the members of the Old Text school put their faith in false or at least dubious documents, they nevertheless tended to be rationalists, enlightened in a sense but not always so favourable to proto-scientific tentatives as the others (one thinks of alchemy, pharmacy, the study of magnetism, etc.). Among the greatest names of the Old Text school were Liu Hsin (50 B.C. to A.D. 23) and Tung Chung-Shu, already met with. By modern times, of course, the attitudes of the ancient scholars to science no longer mattered; what was at stake was the authenticity of the classics. So far as we can now tell, the differences between the 'old' and 'new' versions were numerous, though fairly minor, but the argument in the nineteenth century was that for some of the classics one no longer had the 'new' versions at all, while the 'old' ones had been produced by the Han scholars themselves. Khang Yu-Wei, for his part, led a great campaign in favour of the New Text school. He believed that the whole of the existing *Tso Chuan* and *Chou Li* had been forged by Liu Hsin himself, and that the *Kungyang Chuan* and the *Li Chi* constituted the only reliable avenue through which the true Confucius could be attained. He also believed that Confucius had been a great reformer rather than a conservative. Hence his two books, the *Hsin Hsüeh Wei Ching Khao* (Study of the Forged Works of the Hsin Dynasty) of 1891, and his *Khung-Tzu Kai Chih Khao* (Confucius as a Reformer) of 1897. Although Khang's philological beliefs are no longer tenable today, the connection between them and his faith in a Confucian blessing on the idea of social progress and evolution will now be evident. Which came first in his own development, the philological conclusions or the social philosophy, is not quite clear.

On the 'Old or New Text' controversies see Tjan Tjoe Som (Tsêng Chu-Sen), *Po Hu Thung, the Comprehensive Discussions in the White Tiger Hall* (Leiden, 1949), vol. 1, pp. 137 ff.; Fung (Fêng) Yu-Lan, *A History of Chinese Philosophy* (tr. D. Bodde, Princeton, 1953), vol. 2, pp. 7 ff., 133 ff., 673 ff.; Woo Kang (Wu Khang), *Les Trois Théories Politiques du Tch'oen Ts'ieou* [*Chhun Chhiu*] (Paris, 1932), pp. 186 ff.; C. S. Gardner, op. cit., pp. 9, 56 ff. On Khang Yu-Wei's thought in detail see Hsiao Kung-Chhüan, 'Khang Yu-Wei and Confucianism', *Monumenta Serica*, 1959, **18**, 96.

As for the role of Confucius himself, whether on the whole progressive or reactionary, debate continues actively among scholars both in China and the West. For a sympathetic and rather convincing statement of the former case see H. G. Creel, *Confucius, the Man and the Myth* (New York, 1949, London, 1951).

evolutionary sense, he chose the former as the title of an extraordinary Utopia, the *Ta Thung Shu* (Book of the Great Togetherness),[1] conceived and first drafted in 1884, partly printed in 1913, and not completely printed till 1935. It has been reprinted in Peking as recently as 1956, and an abridged English translation appeared in 1958,[2] so that Western readers now have access to a magnificent description of the future, visionary perhaps but extremely practical and scientific, which it would not be inappropriate to call Wellsian in its authority and scope, a vision which no Chinese scholar could have been expected to create if his intellectual background had been as timeless and static as Chinese thought has only too often been supposed to be. Khang Yu-Wei predicted a supra-national co-operative commonwealth with world-wide institutions, enlightened sexual and racial policies, public ownership of the means of production, and startling scientific and technological advances including the use of atomic energy. In our own time the charismatic phrases of old became the nationwide watchwords of the political parties, *Thien hsia wei kung* (Let the whole world be One Community) for the Kuomintang, and *Thien hsia ta thung* (The world shall be the Great Togetherness) for the Kungchhantang.

Enough has surely now been said to demonstrate conclusively that the culture of China manifested a very sensitive consciousness of time. The Chinese did not live in a timeless dream, fixed in meditation upon the noumenal world. On the contrary, history was for them perhaps more real and vital than for any other comparably ancient people; and whether they conceived time to contain a perennial fall from ancient perfection, or to pass on in cycles of glory and catastrophe, or to testify to a slow but inevitable evolution and progress, time for them brought real and fundamental change. They were far from being a people who 'took no account of time'. And to what extent they often

[1] Note that the title was taken from the *Li Chi* but the progressive content from the traditional development of the *Kungyang Chuan*, and from *Mo Tzu*.

[2] By L. G. Thompson, *Ta Thung Shu: the One-World Philosophy of Khang Yu-Wei* (London, 1958).

visualized it in terms of progress we can see by following another
line of thought.

THE DEIFICATION OF DISCOVERERS AND
THE RECOGNITION OF ANCIENT TECHNOLOGICAL
STAGES IN TIME

No classical literature in any civilization paid more attention to
the recording and honouring of ancient inventors and innovators
than that of the Chinese, and no other culture, perhaps, went so
far in their veritable deification so late in historical times.[1] Texts
which might be termed techno-historical dictionaries, or records
of inventions and discoveries, form a distinct genre.[2] The oldest
one of the kind is the Shih Pên (Book of Origins), most of which
simply recites the names and deeds of the legendary or semi-
legendary culture-heroes and inventors, often dubbed 'ministers'
of the Yellow Emperor, systematizing thus a body of legendary
lore more copious than that of the 'technic deities' of Mediterranean antiquity. Thus Su Sha invented salt-making, Hsi Chung
invented carts and carriages, Chiu Yao invented the ard, Kung-
shu Phan the rotary millstone, and Li Shou computations. Five
or six classes of these names have been distinguished: the clan
patrons and ancestors, the gods of antiquity demoted to heroes,
the patron deities of trades, the mythical heroes euhemerized to
inventors, then certain made-up names of transparent etymology
(like the first example above), and lastly the inventors who were
undoubtedly historical personages, such as the fourth of the above
examples. The history of the text, which we now have in eight
versions, is complicated, but there can be no doubt that while
Ssuma Chhien used one form of it, it never had anything to do
with Master Tsochhiu the historian, to whom third-century A.D.
scholars attributed it. The most probable view is that it was first

[1] The Huai Nan Tzu book points the moral by saying that what rendered the
culture-heroes worthy of divine honours was the outstanding service which they
rendered to the benefit of mankind (ch. 13, Morgan tr., p. 178).
[2] Cf. SCC, Vol. I, pp. 51 ff.

put together by somebody in Chao State between 234 and 228 B.C., thus just a little later than the *Lü Shih Chhun Chhiu*. From the post-Han centuries one could find a dozen or more books to place in this category, and writers were still not tired of it as late as the Ming, when Lo Chhi wrote his *Wu Yuan* (On the Origin of Things) some time in the fifteenth century.

So greatly prized was the lore of the traditional inventors of old that a list of them was incorporated into one of the greatest arcana of Chinese naturalistic philosophy, the *I Ching* (Book of Changes). This is a very strange classic; it took its origin from what was probably a collection of peasant omen texts, accreted a large amount of material concerned with ancient divination practices, and ended as an elaborate system of symbols with their explanations—sixty-four patterns of long and short lines in all possible permutations and combinations. Since to each of these was assigned a particular abstract idea, the whole system played the part of a repository of concepts for developing Chinese science, the symbols being supposed to represent a gamut of forces actually acting in the external world. The continuing additions to the book made by many profound minds through the ages in the form of appendices and commentaries turned it into one of the most remarkable works in all world literature, and gave it immense prestige in traditional Chinese society, so that philosophical sinologists are still today studying it with great interest.[1] One indeed wrote only a few years ago on the concept of time in the *I Ching*, showing how inescapably this is bound up with its theme—'Change, that is the only thing in the universe which is Unchanging'.[2] Nevertheless others have felt that on the whole the *I Ching* exerted an inhibitory effect on the development of the natural sciences in China, since it tempted men to rest in schematic explanations which were not explanations at all. It was in fact a vast filing system for natural novelty, a convenient

---

[1] See H. Wilhelm, *Die Wandlung; acht Vorträge zum I-Ging (I Ching)* (Peking, 1944), Eng. tr. by C. F. Baynes, *Change; Eight Lectures on the I Ching* (London, 1961).

[2] H. Wilhelm, 'Der Zeitbegriff im "Buch der Wandlungen"', *Eranos Jahrbuch*, 1951, **20**, 321.

mental chaise-longue which avoided the need for further observation and experiment.[1]

The dating of the *I Ching* is a highly involved question, but we shall not go far wrong if we place the canonical text (a compilation of omens) mainly in the eighth century B.C., though not complete until the third century B.C., while the principal appended writings (the 'Ten Wings') must date from the Chhin and Han, not finalized until the second century A.D. One of these appendices makes now a curious correlation between the great inventions and a select number of the symbols.[2] Precisely from these, it is alleged, the culture-heroes got their ideas. In other words, the scholars of the Chhin and Han found it necessary to adduce reasons for the inventions from the corpus of symbols in the concept-repository. Nets, textile-weaving, boat-building, houses, the crafts of the archer, the miller and the accountant, all are derived ingeniously from Adherence, Dispersion, Massiveness, Cleavage, the 'Lesser Topheaviness', the 'Break-through', and the like. What this teaches us here is chiefly, I think, the honour that was done to the venerated technic sages by incorporating them in the sublime world-system of the 'Book of Changes'.

There was also more concrete liturgical veneration. Everyone who spends time in China and travels about in the different provinces is deeply impressed by the many beautiful votive temples dedicated not to Taoist gods or to Buddhas and Bodhisattvas but to ordinary men and women who conferred benefits upon posterity. Some keep up the memory of great poets, such as the Tu Fu Tshao Thang at Chhêngtu, others that of great commanders such as the Kuan Kung Ling south of Loyang. But the technicians have a most eminent place. Twice in my lifetime I have had the privilege of burning incense (literally or metaphorically) at Kuanhsien in the temple of Li Ping (*fl.* 309 to 240 B.C.), the great hydraulic engineer and governor of Szechuan province, which stands and has for centuries stood beside the

---

[1] Cf. *SCC*, Vol. II, pp. 336 ff.

[2] R. Wilhelm & C. F. Baynes tr., vol. I, pp. 353 ff. Tabulation in *SCC*, Vol. II, p. 327.

great cutting made under his leadership through the ridge of a mountain. This work divides the Min River into two parts and irrigates still today an area fifty miles square supporting some five million people who can till the soil free of the danger of drought and flood. Every branch of science and technique is represented in these temples of doers and makers deified by popular acclamation. The great physician and alchemist of the Sui and Thang, Sun Ssu-Mo (c. A.D. 601 to 682) has such a temple, and the custom did not cease even in the Ming, for Sung Li (d. A.D. 1422), the engineer who made the summit levels of the Grand Canal a practical proposition, was given a votive temple posthumously beside its very waters.[1] Nor was incense burnt only to men. Huang Tao-Pho (*fl.* A.D. 1296) was a famous woman textile technologist, instrumental in the propagation of cotton growing, spinning and weaving, which she brought to the Yangtze Valley from Hainan. The towns and villages of the cotton areas all honoured her, and built many votive temples to her after her death.[2] It is thus impossible to maintain that the Chinese people had no recognition of technical progress. It may have proceeded at a leisurely rate very different from what we have been accustomed to since the rise of modern science, but the principle is clear.

We can also see it in quite another, and rather unexpected way. The conception of the three major technological stages of man's culture, the ages of stone, bronze and iron following each other in a universal series, has been called the corner-stone of all modern archaeology and prehistory.[3] In its modern form this

---

[1] The Grand Canal, connecting Hangchow in the south with Peking in the north directly, embodied the earliest successful fully artificial summit canal in any civilization, across the foothills of the mountains of Shantung. Planned originally by the astronomer and engineer Kuo Shou-Ching, this section was built by the Mongol military engineer Oqruqči (Ao-Lu-Chhi), with a Chinese colleague Ma Chih-Chên, in A.D. 1287. But it was not capable of full year-round efficiency, however, until Sung Li in A.D. 1411 succeeded in capturing the waters of certain mountain rivers by dams and subsidiary feeder canals, thus ensuring adequate water-levels in the summit sections at all times. For further details see *SCC*, Vol. IV, pt. 3.

[2] Details will be given in *SCC*, Vol. V.

[3] G. Daniel, *The Three Ages* (Cambridge, 1943), p. 9, citing with approval R. A. S. Macalister, *Textbook of European Archaeology* (London, 1921).

was crystallized in 1836 by the Danish archaeologist C. J. Thomsen,[1] who used it to bring some order into the massive collections of the national museum at Copenhagen of which he was director.[2] His good fortune was that during the subsequent decade the generalization was placed for the first time on a fully scientific basis by the stratigraphical excavations of his compatriot J. J. A. Worsaae, also in Denmark.[3] Though sometimes criticized in recent years, it remains the basic classification of periods of high antiquity, and a permanent part of human knowledge. For its general acceptance there were several limiting factors. Thus it had first to be acknowledged that stone tool artifacts had indeed been made by man (and this was only slowly accepted after the Renaissance as acquaintance with existing primitive peoples grew).[4] It was necessary also to understand the correlation of orderly series of geological strata with time, and to escape from the prison of traditional biblical chronology so as to recognize the archaeological evidence of man's true antiquity.[5] Furthermore it was necessary to link archaeological findings with some know-

[1] See his *Ledetrad til Nordiske Oldkindighed* (Copenhagen, 1836), Germ. tr. *Leitfaden zur nordischen Altherskunde* (Copenhagen, 1837), Eng. tr. *A Guide to Northern Antiquities* (London, 1848).

[2] See the stimulating paper by R. F. Heizer, 'The Background of Thomsen's Three-Age System', *Technol. & Cult.* 1962, **3**, 259.

[3] See his *Primaeval Antiquities of Denmark*, tr. from Danish by W. J. Thoms (London, 1849).

[4] The idea that neolithic flaked and polished stone implements were meteorites is old in Europe, whence probably it passed to China by the eighth century A.D., for the pharmaceutical naturalist Chhen Tshang-Chhi (*fl.* A.D. 713 to 733) is the first to speak of 'thunder-axes' (*phi-li fu*), an appellation afterwards common in Chinese scientific literature. Cf. *SCC*, Vol. III, pp. 434, 482, and B. Laufer, *Jade* (Field Museum, Chicago, 1912, repr. Perkins, Pasadena, 1946), pp. 63 ff. The Chinese were always familiar with stone (flint) arrowheads (*shih nu*) and associated them with a distant north-eastern people called the Su-Shen. There is an old story that Confucius gave this explanation when a sparrow-hawk with such an arrowhead in it fell dead in the palace of the Prince of Chhen; he said also that if the treasury was opened ancient examples would be found conserved there, and they were (*Shih Chi*, ch. 47, Chavannes tr., vol. 5, pp. 340 ff. Cf. Laufer, op. cit., pp. 55 ff.). The term Su-Shen may be the oldest Chinese transcription of Jurchen, the Tartar or Tungusic people who later founded the Chin dynasty (A.D. 1115 to 1234) and were related to the Manchus.

[5] Cf. *SCC*, Vol. III, p. 173.

The Grand Titration

ledge of the distribution of metallic ores and some reconstruction of the most primitive techniques of copper, bronze and iron production. Nevertheless Thomsen was only the nucleus of crystallization, for the general idea had been 'in the air' since the middle of the sixteenth century, a time when curious enquirers into 'fossilia' were, as humanists, well acquainted with Greek and Latin texts. They certainly knew of the passage in Lucretius which distinguishes the three ages:

> arma antiqua manus ungues dentesque fuerunt
> et lapides et item silvarum fragmina rami,
> et flamma atque ignes, postquam sunt cognita primum.
> posterius ferri vis est aerisque reperta.
> et prior aeris erat quam ferri cognitus usus,
> quo facilis magis est natura et copia major.[1]

> '... Man's ancient arms
> Were hands and nails and teeth, stones too and boughs
> Broken from forest trees, and flame and fire
> As soon as known. Thereafter force of iron
> And bronze discovered was; but bronze was known
> And used ere iron, since more amenable
> Its nature is and its abundance more.'
> (tr. Leonard, mod.)

This has been called 'just a general scheme of the development of civilization, and based entirely on abstract speculation'.[2] I am not so sure that Lucretius never picked up a flaked arrow-head. At any rate his contemporaries in China were saying exactly the same thing, with no less appreciation of the rise of man in time from primitive savagery, and with perhaps more sure and certain reason for what they averred.

Lucretius' words may have been written in the neighbourhood of 60 B.C. The *Yüeh Chüeh Shu* (Lost Records of the State of

[1] *De Rer. Nat.*, V, 1283 ff.
[2] Daniel, loc., cit., p. 13. These words would be much better applicable to Hesiod's account of five ages (gold, silver, bronze, heroic and iron) in the *Works and Days* (eighth century B.C.) ll. 110 ff., which is really not in the running as a description of actual technological periods, though some have sought to make it so. See J. G. Griffiths, 'Archaeology and Hesiod's Five Ages' in *Journ. History of Ideas*, 1956, **17**, 109, with comment by H. C. Baldry, pp. 553 ff.; F. J. Teggart, 'The Argument of Hesiod's Works and Days', *Journ. History of Ideas*, 1947, **8**, 45.

272

Yüeh), a feudal princedom absorbed by Chhu State in 334 B.C., is attributed to Yuan Khang, a scholar of the Later Han, whose work, which certainly made use of ancient documents and oral tradition, was completed by A.D. 52. Here, in the chapter on the work of the swordsmiths we find the following passage.[1] The Prince of Chhu (Chhu Wang) is engaged in a discussion with an adviser named Fêng Hu Tzu.

'The Prince of Chhu asked: "How is it that iron swords can have the wonderful powers of the famous swords of old?"

Fêng Hu Tzu replied: "Every age has had its special ways (of making things). In the time of Hsien-Yuan, Shen Nung and Ho Hsü weapons were made of stone, (and stone was used for) cutting down trees and building houses, and it was buried with the dead. Such were the directions of the sages. Coming down to the time of Huang Ti, weapons were made of "jade", and it was used also for the other purposes, and for digging the earth; and it too was buried with the dead, for jade is a numinous thing. Such were the directions of the (later) sages. Then when Yü (the Great) was digging (dykes, and managing the waters), weapons were made of bronze. (With tools of bronze) the I-Chhüeh defile was cut open and the Lung-mên gate pierced through; the Yangtze was led and the Yellow River guided so that they poured into the Eastern Sea--thus there was communication everywhere and the whole empire was at peace. (Bronze tools) were also used for building houses and palaces. Was not all this also a sagely accomplishment? Now in our own time iron is used for weapons, so that each of the three armies had to submit, and indeed throughout the world there was none who dared to withhold allegiance (from the High King of the Chou). How great is the power of iron arms! Thus you too, my Prince, possess a sagely virtue."

The Prince of Chhu answered: "I see; thus it must have been." '

[1] Ch. 13. Attention was first drawn to it in the West by Friedrich Hirth in 1904, 'Chinesische Ansichten ü. Bronzetrommeln', in *Mitt. d. Sem. f. Or. Spr.*, **7,** 200 (pp. 215 ff.). This was a brilliant pioneer effort in the history of metallurgy and proto-archaeology. Hirth returned to the periodization question in his *Ancient History of China, to the End of the Chou Dynasty* (New York, 1908, repr. 1923), p. 236. The dating of the *Yüeh Chüeh Shu* and its putative compiler depends on a statement at the end of ch. 3 that in A.D. 52 a period of 567 years had elapsed since a certain event in Yüeh history, fixing that at 515 B.C. But this is simply an appended sentence and cannot in itself date the whole material of the book. From the internal evidence of the style of the passage in question, Hirth was inclined to place it as early as the fifth century B.C., and while this would be almost impossible to prove it is not an entirely unreasonable estimate. It would correspond rather well with what we know of the iron industry at that time. The *Yüeh Chüeh Shu* passage continues to interest archaeologists; cf. Chang Kuang-Chih, *The Archaeology of Ancient China* (New Haven, 1963), p. 2.

Here then, apart from the intercalation of a 'jade' sub-period, possibly meaning stone of better qualities, but also perhaps referring to worked stone as distinct from unworked stone, we have a sequence just as clear as that of Lucretius. And Yuan Khang had a double advantage.

First, he belonged to a distinct tradition.[1] If we read the books of the Warring States philosophers we find time after time a lively appreciation of the stages which mankind had passed through in attaining the high civilization of the late Chou period.[2] The Taoists and Legalists from the fifth century B.C. onwards worked out a highly scientific version of ancient history and social evolution.[3] They had at their disposal the ancient epics of Yao and Shun enshrined in chronicles like the *Chu Shu Chi Nien* (Bamboo Books), a text which came down from the Wei State just as the *Chhun Chhiu* had emanated from Lu, they had the lists of culture-heroes and inventors which ultimately formed the substance of the *Shih Pên*, and they had plenty of oral mythological traditions. From these they made their culture-stage sequence with conscious reference to the customs of the primitive peoples around them. They spoke of men living in nests in trees (pile-dwellings perhaps) or holes in the ground (including cave-dwellings), of the food-gathering stage and the origin of fire and cooked food, of the first making of clothes, the development of the art of the potters (whose neolithic Yangshao and Lungshan wares are now so well-known), and of the first writing on bone and tortoise-shell. A passage in the *Han Fei Tzu* book (*c.* 260 B.C.) relating a speech of Yu Yü to the Prince of Chhin, suggests

[1] 'The Chinese' wrote Hirth (*Bronzetrommeln*, p. 215) 'began to study the developmental periods of prehistory relatively early, and drew their conclusions from tomb finds and other cultural remains.'

[2] It was one of the great merits of my teacher, Gustav Haloun, that he brought out this in his paper 'Die Rekonstruktion der chinesischen Urgeschichte durch die Chinesen', *Japanisch-Deutsche Zeitschr. f. Wiss. u. Tech.* 1925, **3**, 243. It was through my *thung chuang*, Laurence Picken, that I came to appreciate this as it deserved.

[3] See, for example *Chuang Tzu*, ch. 29, *Kuan Tzu*, ch. 84, *Mo Tzu*, ch. 25, *Shang Chün Shu*, ch. 7, *Han Fei Tzu*, ch. 10, *Lü Shih Chhun Chhiu*, ch. 117. Among works of rather later date cf. *Li Chi*, chs. 5, 9, *Lieh Tzu*, ch. 5, *Ho Kuan Tzu*, ch. 13, *Huai Nan Tzu*, ch. 13.

strongly that the writer had seen neolithic pottery both red and black, and also the bronze vessels of the Shang cast in deep relief.[1] Wood, stone, 'jade' (worked stone), bronze and iron, were regularly associated, as in the passage just quoted from the *Yüeh Chüeh Shu*, with one or other of the mythological rulers.[2] A whole book could well be written on this antique 'proto-archaeology'.[3]

Secondly China differed from Europe in that the three technological ages had succeeded each other rather faster, and were thus almost parts of history rather than prehistory. Stone tools were still in general use in the Shang kingdom (fifteenth to eleventh centuries B.C.) and continued so down to the middle of the Chou, probably till the coming of iron, for it seems that bronze was very little used for agricultural tools at any time. It is revealing that the physicians maintained a persistent tradition that in ancient times their acupuncture needles had been sharply pointed pieces of stone (possibly obsidian).[4] The neolithic cultures earlier than the Shang were known under the general name of a Hsia period or 'kingdom', and it was well realized that they had had no bronze. Copper, tin and bronze metallurgy, however, quickly reached great heights of expertise under the Shang, and the 'beautiful metal', as it was called, remained in use for weapons and for marvellous sacrificial and commemorative bronze vessels

[1] Ch. 10, cf. W. K. Liao tr. vol. 1, pp. 85 ff.; Chang Kuang-Chih (loc. cit.). There is a parallel passage in the *Shih Chi*, ch. 5, tr. Chavannes, vol. 2, p. 40 ff. The incident is attributed to 626 B.C. The Prince of Chhin whom Yu Yü addressed was the same Duke Mu who a few years later died and was buried with human sacrifices, 'therefore he had never been able to obtain the hegemony, etc.', see p. 225, above.

[2] Hirth's view (*Ancient History*, pp. 13 ff.) that the legendary Chinese 'emperors' should be regarded as 'symbols of the . . . phases of Chinese civilization' and 'representatives of preparatory periods of culture' is quite acceptable today, and the criticisms of B. Laufer (*Jade*, pp. 70 ff.) were wide of the mark. Haloun (*Rekonstruktion*) believed that they had all been originally cosmological gods of world regions and patron-deities of clans, then gradually they became culture-heroes and at last 'rulers'.

[3] It seems quite strange that no one has collected and studied all the passages in Chou, Chhin and Han literature on remote antiquity from this point of view.

[4] Probably the oldest reference is in the *Shan Hai Ching* (Classic of the Mountains and Rivers), a Chou text, ch. 4, de Rosny tr. p. 158.

down to the middle of the Chou.[1] The introduction of iron then occurred in perfectly historical times, a little before the life of Confucius, towards the middle of the sixth century B.C.,[2] and it is not difficult now to trace many of the profound economic and social effects which it brought about.[3] Those, therefore, who have sought to dismiss Yuan Khang's generalization as cavalierly as that of Lucretius, have had even less justification. 'This is not a case', someone wrote, 'of genius forestalling science by two thousand years; an alert intelligence is simply juggling possibilities without any basis in fact or any attempt to test them.'[4] Actually, neither of these alternatives is applicable. The scholars of the Chou and Han did not make stratigraphical excavations, but they had a far more secure basis for their conviction of the truth of the three technological stages than such a criticism could conceive. For the very tempo of development of their civilization had made them historians rather than prehistorians.

SCIENCE AND KNOWLEDGE AS CO-OPERATIVE ENTERPRISE CUMULATIVE IN TIME

It is possible to follow the conception of a progressive development of knowledge a good deal further, far beyond the level of

[1] Indeed down to the Chhin and Early Han, for Chhin Shih Huang Ti called in all the bronze he could as a disarmament measure after the empire's unification, and had colossal figures made of it (*Shih Chi*, ch. 6).

[2] Cf. J. Needham, 'Remarks on the History of Iron and Steel in China', with French tr. in *Actes du Colloque International Le Fer à travers les Ages*, in *Annales de l'Est* (Nancy) 1956, no. 16, pp. 93, 103; and more fully in 'The Development of Iron and Steel Technology in China', Newcomen Soc. London, 1958 (Dickinson Memorial Lecture), repr. Heffer, Cambridge, 1964, with French tr. (unrevised, with some illustrations omitted and others added), *Revue d'Histoire de Sidérurgie*, 1961, **2**, 187, 235, 1962, **3**, 1, 62.

[3] Cf. Chêng Tê-Khun, *Archaeology in China*, vol. 3, *Chou China* (Toronto, 1963), pp. 246 ff.; Chang Kuang-Chih, op. cit., pp. 195 ff.

A very remarkable feature of Chinese iron technology is the fact, now generally accepted, that cast iron was made almost as soon as iron itself was known, while in the West it took more than seventeen further centuries to achieve this. A number of technical reasons are available in explanation but great credit is due to the ancient Chinese iron-masters.

[4] R. H. Lowie, *The History of Ethnological Theory* (London, 1937), p. 13.

ancient techniques. It would be quite a mistake to imagine that Chinese culture never generated this conception, for one can find textual evidence in every period showing that in spite of their veneration for the sages, Chinese scholars and scientific men believed that there had been progress beyond the knowledge of their distant ancestors.[1] Indeed the whole series of astronomical tables ('calendars', see p. 232 above) illustrates the point, for each new emperor wanted to have a new one, necessarily better and more accurate than any of those that had gone before.[2] No mathematician or astronomer in any Chinese century would have dreamed of denying a continual progress and improvement in the sciences which they professed. How right they were may be appreciated from Fig. 31 which shows the gradual increase in the accuracy of mechanical timekeeping. The same also may be said to be true of the pharmaceutical naturalists, whose descriptions of the kingdoms of Nature grew and grew. In Fig. 32 the number of main entries in the pharmacopoeias between A.D. 200 and 1600 are plotted so as to show the growth of knowledge through the centuries; the unduly sharp rise just after A.D. 1100 is probably referable to increasing acquaintance with foreign, especially Arabic and Persian, minerals, plants and animals, with

[1] The importance of establishing this point arose in correspondence with Mr Arthur Clegg.

[2] One striking example may be given; the saecular variation in the length of the tropical year. Over the ages the Chinese astronomers came to recognize that there had been a very gradual shortening (hsiao chang); this was first computed in A.D. 1194 and confirmed with exceptionally accurate observations in A.D. 1282. The value obtained for this very minute term was actually much too large, probably because it was desired to 'save' three extant observations made in the first millennium B.C. (and probably very inaccurate), but the whole story is a remarkable example of the cumulative and progressive Chinese endeavour to improve gradually upon all existing previously accepted values. It also shows how essentially correct conclusions, much more spacious and open-minded than those which Europeans of those times were willing to entertain, could often be reached though the numerical data were wrong. Proper motion is another example of this (SCC, Vol. III, p. 270). On the saecular variation of the tropical year length, see Nakayama Shigeru, Jap. Journ. Hist. Sci. 1963, 68, 128 and Abstracts of Communications to the Xth. Internat. Congress of the History of Science N.Y., 1962, p. 90, as also Jap. Studies in the Hist. of Sci., 1963, 2, 101.

a synonymic multiplication which subsequently righted itself.[1] The position in China would be well worth contrasting in detail with that in Europe. In his great work Bury showed long ago that before the time of Francis Bacon only very scanty rudiments of the conception of progress are to be found in Western scholarly literature.[2] The birth of this conception was involved in the famous sixteenth- and seventeenth-century A.D. controversy between the supporters of the 'Ancients' and those of the 'Moderns', for the studies of the humanists had made it clear that there

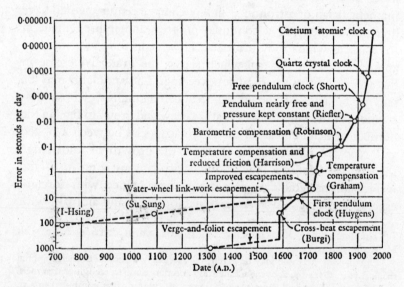

FIGURE 31   Chart to show the rise in accuracy of the mechanical clock through the centuries (amplified by J. Needham from the original graph of F. A. B.Ward, with his approval, after consultation with J. H. Combridge and H. von Bertele).

[1] The data used are taken from Yen Yü, 'Shih-liu Shih-chi-ti Wei Ta Kho-Hsüeh Chia Li Shih-Chen (The Great Sixteenth-Century Scientist Li Shih-Chen)', art. in *Chung-Kuo Kho-Hsüeh Chi-Shu Fa-Ming ho Kho-Hsüeh Chi-Shu Jen Wu Lun Chi* (Chinese Scientific and Technological Discoveries and the Men who Made them), ed. Li Kuang-Pi & Chhien Chün-Yeh, Peking 1955, p. 314; see also Chêng Chih-Fan, 'Li Shih-Chen and his Materia Medica', *China Reconstructs*, 1963, **12**, (no. 3), 29. In SCC, Vol. V, we shall present a full account of the Chinese traditions of pharmaceutical natural history and of Li Shih-Chen's place in it.

[2] J. B. Bury, *The Idea of Progress* (London, 1920).

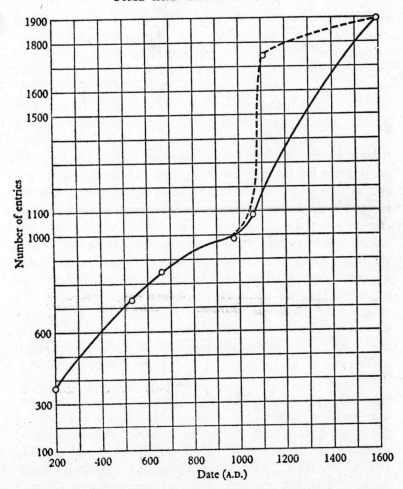

FIGURE 32   Chart showing the increase in the number of entries in the Chinese pharmacopoeias through the centuries (graph by J. Needham based on the census of Yen Yü).

were many new things, such as gunpowder, printing and the magnetic compass, which the ancient Western world had not possessed. The fact that these (and many other innovations) had come from China or other parts of Asia, was long overlooked, but the history of science and technology as we know it was born at the same time out of the perplexity which this discovery had generated.[1] Bury had dealt with progress in relation to the history of culture in general; Zilsel enlarged his method to deal with progress in relation to the ideal of science.[2] The 'ideal of scientific progress' included, he thought, the following ideas (a) that scientific knowledge is built up brick by brick through the contributions of generations of workers, (b) that the building is never completed, and (c) that the scientist's aim is a disinterested contribution to this building, either for its own sake, or for the public benefit, not for fame or private personal advantage. Zilsel was able to show very clearly that expressions of these beliefs, whether in word or deed, were extremely unusual before the Renaissance, and even then they developed not among the scholars, who still sought individualistic personal glory, but among the higher artisanate, where co-operation sprang quite naturally from working conditions. Since the social situation in the era of the rise of capitalism favoured the activities of these men, their ideal was able to make headway in the world. Zilsel traces the first appearance of the idea of the continuous advancement of craftsmanship and science to Mathias Roriczer, whose book on cathedral architecture appeared in A.D. 1486.[3] Thus 'science' said Zilsel, 'both in its theoretical and utilitarian interpretations, came to be regarded as the product of a co-operation for non-personal ends, a co-operation in which all scientists of the past, the present and the future have a part.' Today, he went on, this idea or ideal seems almost self-evident—yet no Brahmanic, Buddhist, Muslim, or Latin scholastic, no Confucian scholar or Renaissance humanist, no philosopher or rhetor of classical antiquity ever achieved it. He would have done better to leave out the reference to the Confucian scholars until Europe knew a little more about them. For in fact it would seem that the idea of cumulative disinterested

[1] Cf. *SCC*, Vol. IV, pt. 2, pp. 6 ff.

[2] E. Zilsel, 'The Genesis of the Concept of Scientific Progress', *Journ. History of Ideas*, 1945, **6**, 325. Cf. also S. Lilley, 'Robert Recorde and the Idea of Progress, a Hypothesis and a Verification', *Renaissance and Mod. Studies*, 1959, **2**, 1.

[3] It was entitled *Von der Fialen Gerechtigkeit* (How to Build Pinnacles and Turrets Correctly), ed. A. Reichensperger (Trier, 1845).

co-operative enterprise in amassing scientific information was much more customary in medieval China than anywhere in the pre-Renaissance West.

There are quotations to be given, but first one must recall that the pursuit of astronomy throughout the ages in China was not the affair of individual star-gazing eccentrics;[1] it was endowed by the State, and the astronomer himself was generally not a freelance but a member of the imperial bureaucracy with an observatory often located in the imperial palace.[2] Doubtless this did harm as well as good, but at any rate the custom of cumulative team-work was deep-rooted in Chinese science. Whole groups of excellent computers and instrument-makers gathered round the great figures such as I-Hsing (A.D. 682 to 727), Shen Kua (A.D. 1031 to 1095) and Kuo Shou-Ching (A.D. 1231 to 1316). And what was true of astronomy was also true of the naturalists, for many of the pharmacopoeias were commissioned by imperial decree, and we know of the large groups who worked together at pharmacognosy and the taxonomic sciences during their twenty-year compiling activities, for example the team led by Su Ching between A.D. 620 and 660. In these respects the medieval scientists of China, building on the knowledge of their forbears, resembled quite closely the historians, who also came together in teams to produce the splendid large-scale works which have already been mentioned (pp. 238 ff.).

Some voices from the past may now give colour to this perhaps unexpected attribute of Chinese culture. Science is cumulative in that every generation builds on the knowledge of Nature acquired by previous generations, but always it looks outward to Nature to see what can be added by empirical observation and new experiment. 'Books and experiments', wrote Edward Bernard in A.D. 1671, 'do well together, but separately they betray an imper-

[1] Naturally this is not my description of the great Greek astronomers, but their contemporaries may well have thought of them in this way; one remembers the anecdote about Thales, recorded by Plato (*Theaet.*, 174 a), that he fell into a well while watching the stars. There may be something else in this (cf. *SCC*, Vol. III, p. 333) but we may take it at its face value here.

[2] Cf. *SCC*, Vol. III, pp. 171 ff., 186 ff.

fection, for the illiterate is anticipated unwillingly by the labours of the ancients, and the man of authors deceived by story instead of science.'[1] This theme of empiricism was extremely strong in the Chinese tradition. 'Those who can manage the dykes and rivers', says the *Shen Tzu* book (probably third century A.D.), 'are the same in all ages; they did not learn their business from Yü the Great, they learnt it from the waters.' 'Those who are good at archery', says the *Kuan Yin Tzu* book (eighth century A.D.), 'learnt from the bow and not from Yi the archer.... Those who can think, learnt from themselves and not from the Sages.'[2] This is in part the message of that splendid story of Pien the wheel-wright in *Chuang Tzu*, who admonished his feudal lord, the Prince of Chhi, for sitting and reading old books instead of learning the art of government from personal knowledge of the nature of people, just as the artisan learns from personal knowledge of the nature of wood and metal.[3] Thus always alongside the Confucian veneration of the sages, and the Taoist threnodies about the lost age of primitive community, there flourished these other convictions that true knowledge had grown and could yet grow immeasurably more if men would look outward to things, and build upon what other men had found reliable in their outward looking. *Ko wu chih chih*—'the attainment of knowledge lies in the investigation of things'—such had been the pregnant phrase in the *Ta Hsüeh* (Great Learning), a book, later one of the classics, written probably by Yochêng Kho, a pupil of Mencius, about 260 B.C., and it was the watchword of Chinese naturalists and scientific thinkers all through the ages.[4]

There is no Chinese century from which one could not cite quotations to illustrate the conception of science as cumulative disinterested co-operative enterprise. Khung Jung (d. A.D. 208) opined, in a passage constantly quoted afterwards, that the ideas of

[1] From S. J. Rigaud, *Correspondence of Scientific Men of the Seventeenth Century* (Oxford, 1841), vol. 1, p. 158, quoted by A. F. Titley, 'Science and History', *History*, 1938, **23**, 108.
[2] Cf. *SCC*, Vol. II, p. 73.
[3] Cf. *SCC*, Vol. II, p. 122.
[4] Cf. *SCC*, Vol. I, p. 48.

intelligent men were often far better for their time than any of the sayings of the ancient sages, and he illustrated his point by a reference to the application of water-wheels to trip-hammer pounding batteries for cereals and minerals.[1] About A.D. 20 already Huan Than had traced the sequence of man-power, animal-power and water-power in industry, a sequence hardly less significant than that of the three technological ages discussed above (pp. 270 ff.).[2] In the field of astronomy and geophysics. Liu Chhuo appealed to the throne in A.D. 604 for the authorization of new research on solar shadow-measurements, proposing the geodetic survey of a meridian arc. He said:

'Thus the heavens and the earth will not be able to conceal their form, and the celestial bodies will be obliged to yield up to us their measurements. We shall excel the glorious sages of old and resolve our remaining doubts (about the universe). We beg your Majesty not to give credence to the worn-out theories of former times, and not to use them.'

His wish was not granted, however, till the following century, when a remarkable 1600-mile meridian arc survey was accomplished between A.D. 723 and 726 under the superintendence of I-Hsing and the Astronomer-Royal, Nankung Yüeh. This did give results different from those previously accepted, and their descriptions show an enlightened recognition that age-old beliefs about the universe must necessarily bow to improved scientific observations, even though the 'scholars of former times' (hsien ju) were discredited thereby.[3] Again, at the end of the eleventh century A.D., the idea of cumulative advance came up against the superstition that each new dynasty or reign-period must 'make all things new', when a new prime minister wanted to destroy the great astronomical clock-tower of Su Sung (cf. pp. 80, 246 above).

[1] Cf. SCC, Vol. IV, pt. 2, p. 392. Here he was but elaborating and applying the doctrine trenchantly stated in the Huai Nan Tau book, ch. 13, tr. Morgan, pp. 143 ff.
[2] SCC, loc. cit.
[3] The whole story is given in SCC, Vol. IV, pt. 1, pp. 44 ff., 53; and more fully in A. Beer, Ho Ping-Yü, Lu Gwei-Djen, J. Needham, E. G. Pulleyblank & G. I. Thompson, 'An Eighth-Century Meridian Line; I-Hsing's Chain of Gnomons and the Prehistory of the Metric System', Vistas in Astronomy, 1961, 4, 3.

There was doubtless an element of party politics here, but two scholar-officials, Chhao Mei-Shu and Lin Tzu-Chung, who warmly admired the clock and regarded it as a great advance on anything of the kind that had previously been made, exerted themselves to pull strings to save it. This they succeeded in doing, and the great clock continued to tick on until the year of doom, A.D. 1126, when the Sung capital was taken by the Jurchen Chin Tartars. They transported it to their own capital, near modern Peking, and re-erected it there, after which it still ran for some decades.[1] It is in connection with these astronomical clocks that we often find the expression 'nothing so remarkable had ever been seen before'. This occurs, for example, in a description of a hydro-mechanical clock with elaborate jackwork constructed under the superintendence of the last Yuan emperor himself, Shun Ti, in A.D. 1354, and though it may be considered a stock literary phrase, it nevertheless reveals the fact that Chinese scholars were very conscious of scientific and technical achievements, by no means always trivial in comparison with the works of the sages of old.[2] It remains to be seen whether, when all the information is in, pre-Renaissance Europe was as conscious of the progressive development of knowledge and technique as they were.

In the light of all this, the widespread Western belief that traditional Chinese culture was static or stagnant turns out to be a typical occidental misconception. But it would be fair to call it 'homeostatic' or 'cybernetic'. For there was something in Chinese society which continually tended to restore it to its original character (that of a bureaucratic feudalism) after all disturbances, whether these were caused by civil wars, foreign invasions, or inventions and discoveries. It is truly striking to see how earth-shaking were the effects of Chinese innovations upon the social systems of Europe, when once they had found their way there, yet they all left Chinese society relatively unmoved. We have mentioned gunpowder, which in the West contributed so powerfully to the over-

[1] Cf. SCC, Vol. IV, pt. 2, pp. 496 ff.; also J. Needham, Wang Ling & Derek J. Price, Heavenly Clockwork (Cambridge, 1960), pp. 116 ff.
[2] Cf. SCC, Vol. IV, pt. 2, p. 507; also Needham, Wang & Price, op. cit., pp. 133 ff.

throw of military aristocratic feudalism, yet after five centuries' use in China left the mandarinate essentially as it had been to start with. At the other extreme, the beginnings of Western feudalism had been associated with the invention of equestrian stirrups, but in China, their original home, no such disturbance of the social order resulted. Or one may take the mastery of iron-casting, achieved in China some thirteen centuries before Europe obtained it— there it was absorbed into customary usage for a great variety of purposes both peaceful and warlike, here it furnished the cannon which destroyed the feudal castle walls, and it formed the machines of the industrial revolution. The fact is that scientific and technological progress in China went on at a slow and steady rate which was totally overtaken by the exponential rise in the West after the birth of modern science at the Renaissance. What is important to realize is that although Chinese society was so self-regulating and stable, the idea of scientific and social progress and real change in time was there. Hence, however great the forces of conservatism there was no ideological barrier of this particular kind to the development of modern natural science and technology when the time was ripe.

TIME AND HISTORY IN CHINA AND THE WEST

We come now to what is perhaps the greatest question of all that have here been raised; could there have been a connection between differences (if any) in the conceptions of time and history characteristic of China and the West and the fact that modern science and technology arose only in the latter civilization? The argument set up by many philosophical writers consists of two parts, first the supposed demonstration that Christian culture was much more historically minded than any other, and secondly the view that this was ideologically favourable to the growth of the modern natural sciences at the Renaissance and the scientific revolution.[1]

---

[1] A conversation with Dr O. Temkin at Baltimore in 1950 was my nucleus of crystallization for this *Fragestellung*, though I had often meditated about it while in China during the Second World War. An interesting fact, and perhaps significant

The first half of the argument has long been familiar ground for occidental philosophers of history.[1] Unlike some other great religions, Christianity was indissolubly tied to time, for the incarnation, which gave meaning and a pattern to the whole of history, occurred at a definite point in time.[2] Moreover Christianity was rooted in Israel, a culture which, with its great prophetic tradition, had always been one for which time was real, and the medium of real change. The Hebrews were the first Westerners to give a value to time, the first to see a theophany, an epiphany, in time's record of events. For Christian thought the whole of history was structured around a centre, a temporal mid-point, the historicity of the life of Christ, and extended from the creation through the *berith* or covenant of Abraham to the *parousia* παρουσια; second coming of Christ), the messianic millennium and the end of the world. Primitive Christianity knew nothing of a timeless God; the eternal is, was, and will be[3] αιωνων των αιωνων, 'unto ages of ages' (in the sonorous words of the Orthodox

in the light of what the following pages contain, is that although I talked about the inhibitory factors of modern science in China with a great number of friends from Dr Kuo Pên-Tao in Chhêngtu to Dr Liang Po-Chiang in Kuangtung, the one thing they never suggested was that the sense of historical time, or the lack of it, could have had anything to do with the matter.

[1] See especially O. Cullmann, *Christus und die Zeit* (Zürich, 1945), Engl tr. *Christ and Time: the Primitive Christian Conception of Time and History*, by F. V. Filson (London, 1951); P. Tillich, *The Interpretation of History*, tr. N. A. Rasetzki & E. L. Talmey (New York & London, 1936); *The Protestant Era*, tr. J. Luther Adams (Chicago, 1948, London, 1951); R. Niebuhr, *Faith and History; a Comparison of Christian and Modern Views of History* (London, 1949); *The Self and the Dramas of History* (London, 1956); H. Christopher Dawson, *The Dynamics of World History*, ed. J. J. Murllo (London, 1957); *Progress and Religion; an Historical Enquiry* (London, 1929); T. F. Driver, *The Sense of History in Greek and Shakespearean Drama* (New York, 1960). Cf. H. Butterfield, op. cit.

[2] It is generally known that the custom of denominating years serially in the A.D. sequence started only from the proposal of Dionysius Exiguus in A.D. 525; it is less well known that the minus series of B.C. years, extending backwards from the birth of Christ, was introduced as late as the seventeenth century A.D., and Bossuet in A.D. 1681 may have been the first to use it. This system is now almost universal; the Chinese speak of *kung yuan, chhien* or *hou*, before or after, in the 'public' or 'international' era, a testimony perhaps to the all-pervadingness of modern technological, rather than specifically Christian, Western civilization. Cf. p. 235, above.

[3] *Rev.*, I, 4.

liturgies); its manifestation the continuous linear redemptive time-process, the plan (*oikonomia*, ὀικονομια) of redemption. In this world-outlook the recurring present was always unique, un-repeatable, decisive, with an open future before it, which could and would be affected by the action of the individual who might assist or hinder the irreversible meaningful directedness of the whole. A moral·purpose in history, the deification of man, was thus affirmed, significance and value were incarnate in it, just as God himself had taken man's nature upon him and died as a symbol of all sacrifice.[1] The world process, in sum, was a divine drama enacted on a single stage, with no repeat performances.

It is customary to contrast this view sharply with that of the Greek and Roman world, especially the former,[2] where cyclical conceptions were generally dominant.[3] We have mentioned the description of successive ages in Hesiod (p. 272), and their eternal recurrence is one of the few doctrines which it is certain that Pythagoras taught;[4] the other end of Hellenism saw the Stoic doctrine of four world periods[5] and the fatalistic pietism of Marcus Aurelius.[6] Eudemus, Aristotle's pupil, envisaged a complete return of time so that once again, or many times again, he would be sitting talking with his students; Aristotle himself,[7] and Plato too,[8] were wont to speculate that every art and science

---

[1] See Irenaeus, *Contra Haeresios*, IV, 37, 7.

[2] Roman thought was rather different, as witness the 'linear' epic of Virgil, the metrical chronicles that preceded it, and the theory of the *urbs aeterna*; cf. Eliade (next note), pp. 201 ff. also C. S. Lewis, 'Historicism', *Month*, 1950 (NS) **4,** 230.

[3] On these see especially M. Eliade, *Le Mythe de l'Eternel Retour; Archétypes et Répétition* (Gallimard, Paris, 1949), Eng. tr. *The Myth of the Eternal Return* by W. R. Trask (London, 1955). Eliade opens his account with a fascinating study of the seasonal rites of ancient and primitive peoples which, he suggests, helped them to protect themselves against the psychological fear of the passage of time and the positive horror of anything really new and irreversible (pp. 80, 128, 184, 217). These repetitions then, joined with the computational long-term cycles of astronomy, led to the Indo-Hellenic universe of cyclical time.

[4] Porphyry, *Vita Pyth.*, 19.

[5] Chrysippus, frgs. 623–627; Zeno, frgs, 98, 109; Eudemus, frg. 51.

[6] *Medit.*, XI. 1.

[7] *Physica*, IV, 14, 223 b 21. *Problemata*, XVII. 3.

[8] *Politics*, 269 c, ff.; *Republic*, VIII, 546.

had many times developed fully and then perished, or that time would return yet again to its beginning and all things be restored to their original state. Such ideas were often combined of course with the long-term recurrences of observational and computational astronomy, hence the notion, probably Babylonian, of the 'Great Year'. Now cyclical recurrence precluded all real novelty, for the future was essentially closed and determined, the present not unique, and all time essentially past time. 'That which has been is that which shall be, and that which has been done is that which shall be done, and there is no new thing under the sun.'[1] Salvation therefore could only be thought of as escape from the world of time, and this was partly what led, as some suppose, to the Greek fascination with the timeless patterns of deductive geometry and the formulation of the theory of Platonic 'ideas'.[2] as well as to the 'mystery-religions'.

Deliverance from the endless repetitions of the wheel of existence at once recalls the world-outlook of Buddhism and Hinduism; and indeed it does seem true that non-Christian Greek thought was extremely like that of India in this respect.[3] A thousand *mahāyugas* (4,000 million years of human reckoning) constituted a single Brahma day, a single *kalpa*; dawning with

[1] *Ecclesiastes*, 1, 9.

[2] Cf. Driver, op. cit., pp. 38 ff. Although historians of science are never tired of hymning the services of Euclidean deductive geometry to the Western world, I vividly remember a conversation with Dr Paul Lorenzen of Bonn in 1949 in which he expressed the view that Europe had had more geometry than was good for it. Of course geometry was an essential basis for modern science, but it did have the bad effect of inducing too ready a belief in abstract timeless axiomatic propositions of all sorts supposedly self-evident, and too willing an acceptance of rigid logical and theological formulations. As these became invested with the authoritarian assurance which the Latin clergy inherited from the Roman jurisconsults, the explosion of the Reformation was inevitable when the merchant class rose to power; and the West still suffers from the slogans of that time. China, however, was algebraic and 'Babylonian', not geometrical and 'Greek', so that opposition tended to be practical and approximative rather than theoretical and absolute, and men did not feel obliged to formulate such timeless axiomatic propositions. Hence empirical, historical, 'statistical' ethics, with little ideological fanaticism and essentially no persecution for religion's sake at all.

[3] Cf. Eliade, op. cit., pp. 167 ff. Belief in the periodical destruction and re-creation of the universe goes back to the tenth century B.C. *Atharvaveda*, X, 8, 39, 40.

re-creation and evolution, ending with dissolution and re-absorption of the world spheres with all their creatures into the absolute.[1] The rise and fall of each *kalpa* brought ever-recurring mythological events,[2] victories of gods and titans alternately, incarnations of Vishnu, churnings of the Milky Ocean to gain the medicine of immortality, and the epic deeds of the *Rāmāyana* and the *Mahābhārata*. Hence the innumerable reincarnations of the Lord Buddha told in the *Jātaka* birth stories.[3] The dimension of the historically unique was not really present in Indian thought, so that India remained by general consent the least historically minded of the great civilizations,[4] while in the Hellenic and Hellenistic situations uninfluenced by Israel, only a few remarkable minds broke through the prevailing doctrine of recurrence, Herodotus and Thucydides, and they but partially. Of course, the hopelessness of this world-outlook was greatly modified in India by the wisdom (more Hindu than Buddhist) of the duty of the householder and husbandman in his generation, in fact its own kind of Stoicism which gave to ordinary social life its honoured place in part at least of every individual's life-cycle.

Paul Tillich has brought together the characteristics of the two

---

[1] Cf. H. Zimmer, *Myths and Symbols*, pp. 11 ff.; 16 ff; 19 ff; Eliade, op. cit., pp. 169 ff.

[2] I think it is important in all this to speak of 'recurrence in time' rather than of 'the reversibility of time' in the cyclical world-theories (though writers on this subject often use the expression), for there was in fact in Buddhist mythology a true doctrine of reversibility, much more far-reaching and much less well known. It has been discussed by P. Mus in his interesting paper already mentioned, 'La Notion de Temps Réversible dans la Mythologie Bouddhique'. As a condition of deliverance from *samsāra*, the flux of becoming, the Time-Ogre, Mahākāla, whom we see in the Tunhuang frescoes devouring all things; the aspiring Bodhisattva had to reverse time's flow, retracing all his previous existences in an inverted order (*pratilomam*). Then, after his last rebirth, he could proclaim himself the 'Firstborn of Time', and stride triumphantly to the summit of the cosmic mountain before disappearing into Nirvāna. This was like running a cinematograph film backwards through the projector, as (with remarkable effect) in certain films of Jean Cocteau.

[3] Eliade, op. cit., pp. 172 ff. A complete translation of these from the Pāli Canon exists: *The Jātaka or Stories of the Buddha's Former Births, translated by various hands,* ed. E. B. Cowell (Cambridge), 6 vols, with index vol., 1895 to 1913.

[4] A good discussion of this will be found in K. Quecke, 'Der Indische Geist und die Geschichte', *Saeculum*, 1950, **1**, 362.

great types of world-outlook into almost epigrammatic form.[1] For the Indo-Hellenic, space predominates over time, for time is cyclical and eternal, so the temporal world is less real than the world of timeless forms, and indeed has no ultimate value.[2] Being must be sought through the fleshly curtain of becoming, salvation can be gained only by the individual, of whom the self-saving *prateyeka buddha* is the prime example, not by the community. The world eras go down to destruction one after the other, and the most appropriate religion is therefore either polytheism (the deification of particular spaces) or pantheism (the deification of all space). It may seem this-worldly, concentrating hedonistically on the passing present, but it dares not to look into the future and seeks lasting value only in the timeless. It is thus essentially pessimistic. For the Judaeo-Christian, on the other hand, time predominates over space, for its movement is directed and meaningful, witnessing an age-long battle between God and evil powers (here ancient Persia joins Israel and Christendom)[3] in which since the good will triumph, the temporal world is ontologically good. True being is immanent in becoming, and salvation is for the community in and through history. The world era is fixed upon a central point which gives meaning to the entire process, overcoming any self-destructive trend, and creating something new which cannot be frustrated by any cycles of time. Hence the most appropriate religion is monotheism, with God as the comptroller of time and all that happens in it. It may seem other-worldly, despising the things of this life, but its faith is tied to the future as well as the past, for the world itself is redeemable, not illusory, and the Kingdom of God will claim it. It is thus essentially optimistic.

We may surely accept, then, as historical fact, the intense history-consciousness of Christendom. The second part of the argu-

---

[1] *Protestant Era*, pp. 23, 30. Cf. Niebuhr, *Faith and History*, pp. 15 ff.

[2] Indeed time is almost absorbed into space, because if every event is infinitely repeated evanescence is only illusory and there is no irreversible change—each moment is like a 'still' from a photographic film which can be run through the projector forwards over and over again (Eliade, p. 184).

[3] Cf. Eliade, op. cit., pp. 185 ff., 191.

ment, which appears to have been hinted at rather than worked out as yet by philosophers of history, is that this consciousness directly contributed to the rise of modern science and technology at the Renaissance, and may therefore rank with other factors in helping to explain it.[1] If it helps to explain it in Europe, perhaps its absence (or putative absence) elsewhere might help to explain the absence of the scientific revolution in those other cultures.

There can be no doubt that time is a basic parameter of all scientific thinking—half of the natural universe indeed, if only a quarter of the number of common-sense dimensions—and therefore that any habit of decrying it cannot be favourable to the natural sciences. It must not be dismissed as illusory; nor depreciated in comparison with the transcendent and the eternal. It lies at the root of all natural knowledge, whether based on observations made at different times, because they involve the uniformity of Nature, or upon experiments, because they necessarily involve a lapse of time, which it may be desirable to measure as accurately

[1] See particularly the interesting recent symposium by F. d'Arcais, A. Buzzati-Traverso, A. C. Jemolo, E. de Martino, Rev. R. Panikkar & U. Spirito, 'Progresso Scientifico e Contesto Culturale', *Civiltà delle Macchine*, 1963, **11** (no. 3), 19. The stimulating but confused book of L. D. del Corral, *El Rapto de Europa; una Interpretación Histórica di Nuestro Tiempo* (Madrid, 1954), Eng. tr. by H. V. Livermore, *The Rape of Europe* (London, New York, 1959), also touches upon this question. So too does the work of K. Jaspers, *Vom Ursprung und Ziel d. Geschichte* (Zürich, 1949), Eng. tr. by M. Bullock, *The Origin and Goal of History* (London, 1953), which recognizes the rise of modern science and technology as the essentially new and unique power-giving component of Western civilization, but accounts for it in ways curiously tentative and faltering for so eminent a thinker. Unlike other presentations of Western philosophy of history mentioned, this is not a Christocentric book, but emphasizes rather the period of the great religious geniuses of all the Old World civilizations *c.* 500 B.C. I may mention also an article by P. F. Douglass, 'Christian Faith and Political Philosophy', *Religion in Life*, 1941, **10**, 267, which surveys many factors possibly connecting the Christian elements in European culture with the rise of modern science in it. To this I was introduced by the late Prof. Roderick Scott in 1944 at Chhangting in Fukien where he was lecturing at the Fukien Christian University evacuated from Fuchow. Prof. Scott dwelt much on the way in which Chinese inventions, while powerless to change society in China, had revolutionized the history of Europe time after time. Long afterwards I went fully into this question in my contribution to *Legacy of China* (Oxford, 1964), but could not adopt the view of Scott and Douglass that Christianity was the main reason for the difference. This is reprinted on pp. 55 ff. above.

as possible.[1] The appreciation of causality, so basic to science, must surely have been favoured by a belief in the reality of time. It is not at first sight obvious, however, why this should have been more favoured by linear Judaeo-Christian rather than by cyclical Indo-Hellenic time, for if the time-cycles were long enough the experimenter would hardly be conscious of them;[2] but it may be that what the recurrence-theories really sapped was the psychology of continuous cumulative never-completed natural knowledge, the ideal that sprang from the higher artisanate but came to fruition in the Royal Society and its virtuosi (cf. p. 280). For if the sum of human scientific effort were to be doomed beforehand to ineluctible dissolution, only to be reformed with endless toil aeon after aeon one might as well seek radical escape in religious meditation or Stoic detachment rather than wearing oneself out like a coral-building polyp engaged with its colleagues in blindly constructing a reef on the rim of a live volcano. Psychological strength was certainly not always weakened in this way, for otherwise Aristotle would never have laboured at his zoological studies, the very title of which is relevant to our thought, Historia Animalium, περι ξωων ιστοριας, showing as it does the original undifferentiated meaning of 'history', any knowledge gained by enquiry—hence the expression still in use, natural history.[3] Nevertheless it is probably reasonable to believe that in sociological terms, for the scientific revolution, where the co-operation of so many men together (unlike the individualism of Greek science) was part of the very essence, a prevalence of cyclical time would have been severely inhibitory, and linear time was the obvious background.

[1] Could this be connected with the fact that the Greeks experimented so little in comparison with their scientific theorizing?

[2] Moreover in a fully cyclical-time matrix there was the doctrine of karma, of the automatic recompense of good or evil deeds in kind, covering many successive reincarnations of the individual. This was (and is) fundamental to Buddhists, whatever philosophical school they might adhere to in the matter of time's reality. As already pointed out (SCC, Vol. II, pp. 418 ff.), however, the law of karma had no stimulating effect on the idea of causality in science, or of Laws of Nature, presumably because the Buddhists were really interested only in the moral part of it, while scientific causality has to be ethically neutral.

[3] This was the kind of personal knowledge of the chroniclers, too, the rerum gestarum scriptores.

Sociologically it may have acted in still another way, it may well have strengthened the resolution of those who worked for a 'root-and-branch reformation in Church and State', bringing into being thereby not only the 'new, or experimental, science' but also the new order of capitalism. Must not the early reformers and merchants alike have believed in the possibility of revolutionary, decisive, and irreversible transformations of society? Linear time could not of course have been one of the fundamental economic conditions which made this possible, but it may have been one of the psychological factors which assisted the process. Change itself had divine authority, no less, for the new covenant had superseded the old, the prophecies had been fulfilled, and with the ferment of the Reformation, backed by the traditions of all the Christian revolutionaries from the Donatists to the Hussites, people dreamt again apocalyptically of the foundation of the Kingdom of God on earth. Cyclical time could contain no apocalyptic. In many ways the scientific revolution, however sober, however patronized by princes, had kinship with these visions; 'That discouraging maxime, *nil dictum quod non dictum prius*', wrote Joseph Glanvill in A.D. 1661, 'hath little room in my estimation; I cannot tye up my belief to the letter of Solomon; these last ages have shown us what antiquity never saw, no, not in a dream.'[1] Perfection no longer lay in the past, books and old authors were laid aside, and instead of spinning cobwebs of ratiocination men turned to Nature with the new technique of mathematized hypotheses, for the method of discovery itself had been discovered. As the centuries passed linear time influenced modern natural science more deeply still, for it was found that the universe of the stars itself had had a history, and cosmic evolution was explored as the background to biological and social evolution.[2]

[1] *Scepsis Scientifica; or, Confest Ignorance the Way to Science, in an Essay on the Vanity of Dogmatising and Confident Opinion* (London, 1661, 1665). Repr. and ed. J. Owen (London, 1885).

[2] An interesting glimpse of two phases of this process may be had in W. Baron & B. Sticker, 'Ansätze z. historischen Denkweise in d. Naturforschung an der Wende vom 18 zum 19 Jahrhundert; I, Die Anschauungen Johann Friedrich Blumenbachs über die Geschichtlichkeit der Natur; II, Die Konzeption der Entwicklung von Sternen und Sternesystemen durch Wilhelm Herschel', *Archiv. f. Geschichte d. Medizin u. d. Naturwissenschaften*, 1963, **47**, 19.

Then the Enlightenment secularized Judaeo-Christian time in the interests of the belief in progress which is still with us, so that although today when 'humanists' or Marxists dispute with theologians they wear coats of different colours, the coats (to an Indian spectator, at least) are actually the same coats, worn inside out.

This brings us to consider the position of Chinese civilization; where did it stand in the contrast between linear irreversible time and the 'myth of eternal recurrence'? There can be no doubt that it had elements of both conceptions, but broadly speaking, and in spite of anything that has been said above, linearity, in my opinion, dominated. Of course European culture was also an amalgam, for although the Judaeo-Christian attitude was certainly dominant the Indo-Hellenic one never died out—one can see this in the Spenglerian view of history in our own time[1] and it has always been so. While Aurelius Augustinus (St. Augustine, A.D. 354 to 430) worked out the Christian system of one-way time and history in his *City of God*,[2] Clement of Alexandria (c. A.D. 150 to 220), Minucius Felix (fl. A.D. 175) and Arnobius (fl. A.D. 300) were inclined to favour astral cycles like the *annus magnus*, the 'Great Year'. Similarly, in the twelfth and thirteenth centuries, just when Joachim of Floris (A.D. 1145 to 1202) was setting forth his evolutionary and apocalyptic theory of the Three Ages, inspired successively by the Three Persons of the Trinity, in his *Liber Introductorius ad Evangelium Aeternum*,[3] Bartholomaeus Anglicus (c. A.D. 1230), Siger of Brabant (A.D. 1277) and Pietro d'Abano (d. A.D. 1316) were prepared at least to discuss with calm, if not complete approbation, the theory that after 36,000 solar years

[1] Cf. also the current debates among radio-astronomers and others, causing no little stir in Cambridge and elsewhere, concerning the 'steady-state' theory of the universe as against the 'expansion and contraction' theory with its corollary of creation and re-creation. The 'Great Year' conception is by no means dead, though it may be hardly recognizable in its scientific overalls.

[2] Cf. also *Confessions*, XI.

[3] Perhaps Joachim's greatest disciple was William Blake, 600 years afterwards, who received the mystical apocalyptic 'antinomian' tradition by way of the Anabaptists, Brethren of the Free Spirit, 'Ranters', and other transmitters of revolutionary Christianity. See the interesting study of A. L. Morton, *The Everlasting Gospel; a Study in the Sources of William Blake* (London, 1958).

history will repeat itself down to the minutest detail owing to the resumption by the planets and constellations of their original places.[1]

For China the case is very similar. Cyclical time was certainly prominent among the early Taoist speculative philosophers (p. 227), in later Taoist religion with its recurring judgment-days (p. 263), and in Neo-Confucianism with its cosmic, biological and social evolution ever renewed after the periodical 'nights' of chaos (p. 251). The second and the third were undoubtedly influenced by Indian Buddhism which brought to China the lore of *mahā-yugas*, *kalpas*, *mahākalpas*, etc., but for this the first was too early, and indeed we do not find in it any developed form of the doctrine, rather a poetic ataraxy based on acceptance of the cyclism of the seasons and the life-spans of living things.[2] But all this leaves out of account both the mass of the Chinese people throughout the ages and also the Confucian scholars who staffed the bureaucracy, assisted the emperor in the rites of the age-old 'cosmism' or Nature-worship, and provided the personnel for the Bureaux of Astronomy and Historiography.[3] Sinologists have appreciated for more than a century the linear time-consciousness of Chinese culture, but whatever they know takes at least as long as that to become the common property of occidental intellectuals.[4] Thus,

[1] See here L. Thorndike, *A History of Magic and Experimental Science during the First Thirteen Centuries of our Era* (New York, 1947 ed.), vol 2, pp. 203, 370, 418, 589, 710, 745, 895.

[2] An example of the persistence of resignation to the cyclical 'wheel of fate', prosperity and decay (*shêng shuai*), may be seen in the conversation between the Tao-Kuang emperor, (r. 1821 to 1850) and one of his high officials Pi Kuei, translated and reported by Meadows, op. cit., pp. 123 ff., 130, 134. 'Alas, in all affairs prosperity is followed by decay!' the emperor kept on repeating.

[3] Pointing up the contrast between the two disciplines, Liu Chih-Chi (cf. p. 237, above) went so far as to advise the exclusion of the astronomical monograph from the dynastic history pattern, presumably because it dealt so much with 'timeless', 'unhistorical', recurring cycles. He was in favour of natural history, however; cf. Pulleyblank, loc. cit., p. 145. A curious parellel will be found in *SCC*, Vol. III, p. 634.

[4] Only someone quite unconversant with Chinese traditions could have written, as C. Dawson did, that the 'denial of the significance of history is the rule rather than the exception among philosophers and religious teachers throughout the ages from India to Greece and from China to Northern Europe' (*Dynamics of World History*, p. 271).

in an interesting paper, Bodde wrote:[1] 'Connected with their intense preoccupation with human affairs is the Chinese feeling for time—the feeling that human affairs should be fitted into a temporal framework. The result has been the accumulation of a tremendous and unbroken body of historical literature extending over more than three thousand years. This history has served a distinctly moral purpose, since by studying the past one might learn how to conduct oneself in the present and future. . . . This temporal mindedness of the Chinese marks another sharp distinction between them and the Hindus.' To the great historical tradition of China we have already referred. It envisaged love (*jen*) and righteousness (*i*) incarnate in human history, and it sought to preserve the records of their manifestation in human affairs. Its 'praise-and-blame' (*pao pien*) bias, 'for aid in government', though somewhat of a limitation and liable to crystallize into dead convention, had nothing to do with the *karma* of Buddhist faith. What it affirmed was that evil social results would follow evil social actions, and though these might lead to the personal ruin of an evil ruler, the effects might also, or only, be visited on his house or dynasty; but inescapable effects there would be. To this the system of rewards and penalties for good or evil actions, worked out through a series of reincarnations of a particular individual, was quite foreign, for the Confucian historians were much more concerned with the community than with the individual. If their time had not been linear it is hardly conceivable that they would have worked with such historical-mindedness and such bee-like industry. Moreover, we have seen that theories of social evolution, technological ages initiated by inventive culture-heroes, and appreciations of the cumulative growth of human science pure and applied, are in no way missing from Chinese culture.

It would easily be possible, finally, to over-estimate comparatively the Judaeo-Christian keying of time's flow to a particular point in space-time when an event of world significance occurred.

[1] D. Bodde, 'Dominant Ideas [in Chinese Culture]', art. in *China*, ed. H. F. McNair (Berkeley & Los Angeles, 1946), pp. 18 ff., 23.

The first unification of the empire by Chhin Shih Huang Ti in 221 B.C. was a never-to-be-forgotten focal point in Chinese historical thinking, all the more important because of the unity of secular and sacred which no schizophrenia of pope and emperor ever broke up. If one demands something still more numinous, the life of the Sage, the Teacher of Ten Thousand Generations, Confucius (Khung Chhiu, 552 to 479 B.C.), supreme ethical moulder of Chinese civilization, the uncrowned emperor, whose influence is vitally alive today in the tenements of Singapore as well as the communes of Shantung, forming the inescapable background of the Chinese mind whether traditional, technical or Marxist; this life was at least as historical as that of Jesus. That the Confucian outlook was essentially backward-looking is a thesis which in the light of the evidence brought forward in this essay alone cannot stand a close look—the Sage's Tao was not put into practice in his own generation but his assurance was that men and women could and would live in peace and harmony whenever and wherever it was practised. When this faith, less other-worldly than Christianity (for Thien Tao, Heaven's Way, was not strictly speaking supernatural), joined with the revolutionary ideas implicit in Taoist primitivism, the radically apocalyptic dreams of Ta Thung and Thai Phing, dreams that men could, and did, fight for, began to exert their potent influences. Tillich wrote: 'The present is a consequence of the past but not at all an anticipation of the future. In Chinese literature there are fine records of the past but no expectations of the future.'[1] Once again it would have been better not to come to conclusions about Chinese culture while Europeans still knew so little about it. The apocalyptic, almost the messianic, often the evolutionary and (in its own way) the progressive, certainly the temporally linear, these elements were always there, spontaneously and independently developing since the time of the Shang kingdom, and in spite of all that the Chinese found out or imagined about cycles, celestial or terrestrial, these were the elements that dominated the thought of the Confucian scholars and the Taoist peasant-farmers. Strange as it may seem

[1] *Protestant Era*, p. 19.

to those who still think in terms of the 'timeless Orient', on the whole China was a culture more of the Irano-Judaeo-Christian type than the Indo-Hellenic.

The conclusion springs to the mind. If Chinese civilization did not spontaneously develop modern natural science as Western Europe did (though much more advanced in the fifteen pre-Renaissance centuries) it was nothing to do with her attidude towards time. Other ideological factors, of course, remain for scrutiny, apart from the concrete geographical, social and economic conditions and structures, which may yet suffice to bear the main burden of the explanation.

8

HUMAN LAW
AND THE LAWS OF NATURE

Hobhouse Lecture, Bedford College, London, 1951; first published in
Journ. History of Ideas, 1951, **12**, 3, 194; revised for Hatfield College
of Technology Lectures, 1961, repr. Mukerji Presentation Volume
(Delhi, 1967).

Without doubt one of the oldest notions of Western civilization
was that just as earthly imperial lawgivers enacted codes of positive
law, to be obeyed by men, so also the celestial and supreme rational
creator deity had laid down a series of laws which must be obeyed
by minerals, crystals, plants, animals, and the stars in their courses.
This idea, we know, was intimately bound up with the development of modern science at the Renaissance in the West. Could its
absence elsewhere have been one of the reasons why modern
science arose only in Europe? In other words, were Laws of Nature
necessary?

If one turns to the best books and monographs on the history of
science, asking the simple question, when in European or Islamic
history was the first use of the term 'Laws of Nature' in the modern
scientific sense, it is hard to find an answer. By the eighteenth
century it was, of course, current coin—most Europeans are
acquainted with these Newtonian words of 1796:

> 'Praise the Lord, for he hath spoken,
> Worlds his mighty voice obeyed;
> Laws, which never shall be broken,
> For their guidance he hath made.'

But this could not, in fact, have been written by a Chinese classical
scholar of the autochthonous tradition. Why?

The will of the lawgiver could embody in codes of enacted statutes, not only laws which had as their basis the immemorial customs of the folk, but also laws which seemed good to him for the greater welfare of the State (or the greater power of the governing class) and which might have no basis in *mores* or ethics. This 'positive' law partook of the nature of the command of an earthly ruler, obedience was an obligation, and precisely specified sanctions followed transgression. This is undoubtedly represented in Chinese thought by the term *fa*, just as the customs of society based on ethics (for example that men do not normally, and should not, murder their parents), or on ancient tabus (for example, incest), are represented by *li*[a], a term which, however, includes in addition all kinds of ceremonial and sacrificial observance.

We learn further that in Roman law two parts were recognized: on the one hand the civil coded law of a specific people or State (positive law), *lex legale*, in the later phrase; and on the other hand the Law of Nations (*jus gentium*), more or less equivalent to natural law (*jus naturale*). The *jus gentium* was presumed to follow the *jus naturale* if the contrary did not appear. Their identity was assumed in Roman law, though not very safely; for (*a*) some customs would certainly not be self-evident to natural reason, and (*b*) there were rules which deserved to be recognized by all mankind, but in fact were not (for example the undesirability of slavery). The traditional origin of this 'natural law' was the increasing residence at Rome of merchants and other foreigners, who were not citizens and therefore not subject to Roman law, and who wished to be judged by their own laws—the best that the Roman jurisconsults could do was to take a kind of lowest common denominator of the usages of all known peoples, and so attempt to codify what would seem nearest to justice to the greatest number of people. Thus it was that the conception of natural law originated.

Natural law was thus the mean of what all men everywhere felt to be naturally right, and 'there came a time', as Maine says, 'when from an ignoble appendage of the *jus civile*, the *jus gentium* came to be considered a great, though as yet imperfectly developed, model to which all law ought as far as possible to conform'.

The distinction is found in Aristotle, who speaks of positive law as δίκαιον νομικόν and of natural law as δίκαιον φυσικόν. He says:

'Political justice is of two kinds, one natural (φυσικόν) and the other conventional (νομικόν). A rule of justice is natural when it has the same validity everywhere, and does not depend on our accepting it or not. A rule of justice is conventional when in the first instance it may be settled in one way or the other indifferently —though having once been settled it is not indifferent; for example, that the ransom for a prisoner shall be a mina, or that a sacrifice shall consist of a goat and not of two sheep, etc. . . . Some people think that all rules of justice are merely conventional, because whereas nature is immutable and has the same validity everywhere, as fire burns both here and in Persia, rules of justice are seen to vary. That rules of justice vary is not absolutely true, but only with qualifications. . . . But nevertheless there is such a thing as natural justice as well as justice not ordained by nature, and it is easy to see which rules of justice, though not absolute, are natural, and which are not natural but legal and conventional, both sorts alike being variable.'   (tr. Rackham)

The passage is very interesting, for it refers to the fact that quantitative and ethically indifferent matters can only be settled by positive legislation, and also trembles on the verge of speaking of Laws of Nature in the scientific sense. Now in the Chinese context there could hardly be a *jus gentium*, for owing to the 'isolation' of Chinese civilization there were no other *gentes* from whose practices an actual universal law of nations could be deduced, but there was certainly a natural law, namely, that body of customs which the sage-kings and the people had always accepted, i.e. what the Confucians called *li*[a].

There can be little doubt that the conception of a celestial lawgiver 'legislating' for non-human natural phenomena has its first origin among the Babylonians. Jastrow gives the translation of Tablet No. 7 of the Later Babylonian Creation Poem, in which the sun-god Marduk (raised to a position of central importance contemporaneously with the unification and centralization under Hammurabi about 2000 B.C.) is pictured as the law-giver to the stars. He it is 'who prescribes the laws for (the star-gods) Anu, Enlil (and Ea), and who fixes their bounds'. He it is who 'maintains the stars in their paths' by giving 'commands' and 'decrees'.

The pre-Socratic philosophers of Greece speak much of necessity (ἀνάγκη), but not of law (νόμος) in Nature. But 'the Sun', Heraclitus says (c. 500 B.C.), 'will not transgress his measures; otherwise the Erinyes, the bailiffs of Diké (the goddess of justice) will find him out'. Here the regularity is accepted as an obvious empirical fact, but the idea of law is present, since sanctions are mentioned. Anaximander, too (c. 560 B.C.), speaks of the forces of Nature 'paying fines and penalties to each other'. But the conception of Zeus Nomothetes in the older Greek poets pictures him as giving laws to gods and men, not to the processes of Nature, for he himself was not truly a Creator. Demosthenes, however (384 to 322 B.C.; living thus between the generation of Mo Ti and that of Mencius), uses the word 'law' in its most general sense when he says: 'Since also the whole world, and things divine, and what we call the seasons, appear, if we may trust what we see, to be regulated by Law and Order'.

Nevertheless, Aristotle never used the law-metaphor, though, as we have noted, he occasionally comes within an inch of doing so. Plato uses it only once, in the *Timaeus*, where he says that when a person is sick, the blood picks up the components of food 'contrary to the laws of nature' (παρὰ τοὺς τῆς φύσεως νόμους). But the conception of the governance of the whole world by law seems to be peculiarly Stoic. Most of the thinkers of this school maintained that Zeus (immanent in the world) was nothing else but κοινὸς νόμος—Universal Law; for example Zeno (*fl.* 320 B.C.); Cleanthes (*fl.* 240 B.C.); Chrysippus; d. 206 B.C.); Diogenes (d. 150 B.C.). It seems more than likely that this new and more definite conception was derived from Babylonian influences, since we know that about 300 B.C. astrologers and star-clerks from Mesopotamia began to spread through the Mediterranean world. Among these one of the most famous was Berossus, a Chaldean who settled in the Greek island of Cos in 280 B.C. Zilsel, alert for concomitant social phenomena, notes that just as the original Babylonian conceptions of Laws of Nature had arisen in a highly centralized oriental monarchy, so in the time of the Stoics, a period of rising monarchies, it would have been natural to view the universe as a great empire, ruled by a divine Logos.

Since, as is known, the Stoic influence at Rome was great, it was inevitable that these very broad conceptions should have their effects in the development of the idea of a natural law common to all men whatever might be their cultures and local customs. Cicero (106 to 43 B.C.), of course, reflects this, saying: *Naturalem legem divinam esse censet (Zeno), eamque vim obtinere recta imperantem prohibentemque contraria*, and elsewhere: 'The universe obeys God, seas and land obey the universe, and human life is subject to the decrees of the Supreme Law.'

Curiously, it is in Ovid (45 B.C. to A.D. 17) that we find the clearest statements of the existence of laws in the non-human world. He does not hesitate to use the word *lex* for astronomical motions. Speaking of the teaching of Pythagoras, he says:

> *in medium discenda dabat, coetusque silentum*
> *dictaque mirantum magni primordia mundi*
> *et rerum causas, et quid natura docebat,*
> *quid deus, unde nives, quae fulminis esset origo,*
> *Juppiter an venti discussa nube tonarent,*
> *quid quateret terras, qua sidera lege mearent,*
> *et quodcumque latet. . . .*

Most translators have failed to do justice to this remarkable statement; Dryden turned it thus:

> 'What shook the stedfast earth, and whence begun
> The Dance of Planets round the radiant Sun . . .'

while King simply left the phrase out altogether. Elsewhere Ovid, complaining of the faithlessness of a friend, says that it is monstrous enough to make the sun go backward, rivers flow uphill, and 'all things proceed reversing Nature's laws (*naturae praepostera legibus ibunt*).'

Far more certain as another contributory line of thought was that which emanated from (or was transmitted from the Babylonians by) the Hebrews. The idea of a body of laws laid down by a transcendent God and covering the actions both of men and of the rest of Nature is frequently met with, as Singer and many others have pointed out. Indeed, the divine lawgiver was one of the most central ideas of Judaism. It would be difficult to overestimate the effect on all occidental thinking in the Christian era of these ideas in the Hebrew scriptures—'The Lord gave his decree to the sea, that the waters should not pass his commandment'. Furthermore, the Jews developed a kind of natural law applying to all men as such, somewhat analogous to the *jus gentium* of Roman Law, in the 'Seven Commandments for the Descendants of Noah'.

Christian theologians and philosophers naturally continued the Hebrew conceptions of a divine lawgiver. In the early centuries of Christianity statements in which laws of non-human Nature are implicit are not difficult to find. For example, the oratorical apologist Arnobius (*c*. A.D. 300), arguing that Christianity is nothing monstrous, says that since its introduction there have been no changes in 'the laws initially established'. The (Aristotelian) elements have not changed their properties. The structure of the machine of the universe (presumably the astronomical system) has not dissolved. The rotation of the firmament, the rising and setting of stars, has not altered. The sun has not cooled. The changes of the moon, the turn of the seasons, the succession of long and short days, have neither been stopped nor disturbed.

We are still in the stage, however, before a sharp separation between (human) natural law and (non-human) Laws of Nature has come about. In the early centuries of the Christian era there are two statements of particular interest which show the ideas in their more or less undifferentiated state. In the *Constitution* of Theodosius, Arcadius, and Honorius of A.D. 395 there is a passage forbidding anyone to practise augury on pain of punishment for high treason: *Sufficit ad criminis molem naturae ipsius leges velle rescindere, inlicita perscrutari, occulta recludere, interdicta temptare*—it is impious to tamper with the principles which keep the secret Laws of Nature from men's eyes. This is strikingly similar to the prohibition of the Chhan-Wei books of augury in China, but here its interest is that it suggests the existence of laws of Nature, connected with the course of human affairs, but not concerned with morality. An interesting parallel concerns the great Chinese thinker Tung Chung-Shu, who in 135 B.C. 'privately analysed the significance of two inauspicious fires'; for this he was sentenced to death, but later amnestied.

The second statement is a famous one of Ulpian, the eminent Roman jurist (d. A.D. 228) whose work occupies so large a part of the Justinian *Corpus Juris Civilis* of A.D. 534.

Jus naturale (he says in the first paragraph of the *Digest*) *est quod natura omnia animalia docuit* . . .
'Natural law is that which all animals have been taught by Nature; this law is not peculiar to the human species, it is common to all animals which are produced on land or sea, and to fowls of the air as well. From it comes the union of man and woman called by us matrimony, and therewith the procreation and rearing of children; we find in fact that animals in general, the very wild beasts, are marked by acquaintance with this law.'   (tr. Monro)

Historians of jurisprudence are at pains to explain that this never had any influence on subsequent legal thinking. This may well be the case, but it was accepted by medieval writers and commentators, and clearly expresses the idea of animals as quasi-'juristic' individuals obeying a code of laws laid down by God. At this point we are very close to the idea of the Laws of Nature as the divine legislation which matter (including animal life) obeys.

As the Christian centuries went on it was inevitable that natural law should come to be identified with Christian morality. St. Paul had clearly expressed it. St Chrysostom (early fifth century) had seen in the Hebrew ten commandments a codification of natural law, and with the *Decretum* of Franciscus Gratianus (A.D. 1148) the identification, never afterwards departed from by orthodox canonists, was complete. It was moreover, as Pollock says, the universal medieval belief that commands of princes contrary to natural law were not binding on their subjects, and could therefore lawfully be resisted. This doctrine, summarized in the phrase *Positiva lex est infra principantem sicut lex naturalis est supra*, bore much fruit at the time of the rise of Protestantism, and the 'right of rebellion against un-Christian princes' had no small part to play in the beginnings of modern European democracy (Gooch). It is interesting to note how precisely it corresponds with the Confucian doctrine, expressed in Mencius, that subjects have a right to dethrone the ruler who ceases to act according to *li*[a]; and the similarity was

certainly not lost upon European social thinkers who read the
Jesuit Latin translations of the Chinese classics after A.D. 1600.

But what of the scientists and their Laws of Nature? We are now
in the seventeenth century, and with Boyle and Newton the concept of Laws of Nature, 'obeyed' by chemical substances and planets
alike, is fully developed. Very little investigation has been made,
however, of the exact points at which it differentiated from the
synthesis of the schoolmen. The lexicographers say that the first
use of the expression in its scientific sense occurs in the first volume
of the *Philosophical Transactions* of the Royal Society (1665).
Thirty years later Dryden inserts it gratuitously in his translation of
the 'Felix qui potuit rerum cognoscere causas' of Virgil's *Georgics*—
it has become a commonplace. Robson, in his excellent book
Civilization and the Growth of Law, regarded it as a specifically
seventeenth-century idea, present in the philosophies of Spinoza
and Descartes as well as in the 'new, or experimental, philosophy'
of the natural scientists. It is the merit of Zilsel to have disentangled
very clearly the stages through which the idea at last came into its
own. We find also in jurists such as Huntington Cairns a recognition of the parallel development in the seventeenth century of
secularized natural law based on human reason, and the mathematical expression of empirical Laws of Nature.

There is no doubt that the turning-point occurs between Copernicus (1473 to 1543) and Kepler (1571 to 1630). The former speaks
of symmetries, harmonies, motions, but never in any place of
laws. Gilbert in his *De Magnete* (1600) does not speak of laws
either, though he enunciates certain generalizations about
magnetism for which the term would have been most suitable.
Francis Bacon's position is complex; in the *Advancement of Learning*
(1605) he speaks of the 'Summary Law of Nature' as the highest
possible knowledge, but doubts whether it can be attained by man;
while in the *Novum Organum* (1620) he uses the term 'law' as
synonymous with Aristotelian substantial form. He had thus really
advanced no further than the scholastics. Galileo, like Copernicus,
never uses the expression 'Laws of Nature', whether in his
Jugendarbeit on mechanism of 1598 or in his *Discourses and Mathe-*

matical Demonstrations on Two New Sciences (1638), which was the beginning of modern mechanics and mathematical physics. What would later have been called laws appear as 'proportions', 'ratios', 'principles', etc. The same remarks apply both to Simon Stevin (whose works date as of 1585 and 1608) and to Pascal (1663); the law-metaphor was not used by them.

By a remarkable paradox, Kepler, who discovered the three empirical laws of the planetary orbits, one of the first occasions on which the Laws of Nature were expressed in mathematical terms, never himself spoke of them as laws, though he used the phrase in other connections. Kepler's first and second 'laws', given in the *Astronomia Nova* of 1609, are paraphrased in long expositions; the third, published in *Harmonices Mundi* (1619), is called a 'theorem'. Yet he speaks of 'law' in connection with the principles of the lever, and in general uses the word as if it were synonymous with measure or proportion.

Since Laws of Nature played so large a part in the astronomical sciences, it has been natural to search mostly among the Renaissance astronomers for the first mentions of them. But a very early reference occurs in connection with quite another group of sciences, geology, metallurgy and chemistry. In his *De Ortu et Causis Subterraneorum* of 1546, Georgius Agricola, discussing the Aristotelian theory of the participation of the element Water in the composition of metals, wrote:

'But what proportion of Earth is in each liquid from which a metal is made, no mortal can ever find out or still less explain, but only the one God has known it, who has given sure and fixed laws to Nature for mixing and blending things together.'

It seems worthy of note that this conception should have emerged at least as early in metallurgical chemistry as in astronomy. Another early formulation occurs in the writings of Giordano Bruno (1548 to 1600), who says that God is to be sought for *in inviolabili intemerabilique naturae lege* (*De Immenso*). But Bruno was very 'Chinese' in his thinking, and showed much more appreciation of the organic character of natural phenomena than most of his European contemporaries.

Meanwhile, an important step in the clarification of the concept had been made by the Spanish theologian Suarez, who in his *Tractatus de Legibus* (1612) made a sharp distinction between the world of morality and the world of non-human Nature, maintaining that the idea of law applied only to the former. He opposed the Thomistic synthesis because it disregarded this distinction. 'Things lacking reason', he says, 'are, properly speaking, capable neither of law nor of obedience. In this the efficacy of divine power and natural necessity . . . are called law by a *metaphor*.' This was clear thinking, and reminds us of the difficulty which the Chinese had in extending the concepts of *li^a* and *fa* to the non-human world. For these same reasons, the introduction of the idea of Laws of Nature to China after the time of the Jesuit Mission met with little response. In an interesting passage written in 1737 d'Argens remarked:

'The Chinese atheists, says a missionary, are not more tractable with relation to Providence, than with regard to the Creation. When we teach them that God, who created the universe out of nothing, governs it by general Laws, worthy of his infinite Wisdom, and to which all creatures conform with a wonderful regularity, they say, that these are high-sounding words to which they can affix no idea, and which do not at all enlighten their understanding. As for what we call laws, answer they, we comprehend an Order established by a Legislator, who has the power to enjoin them, to creatures capable of executing these laws, and consequently capable of knowing and understanding them. If you say that God has established Laws, to be executed by Beings capable of knowing them, it follows that animals, plants, and in general all bodies which act conformable to these Universal Laws, have a knowledge of them, and consequently that they are endowed with understanding, which is absurd.'

In Descartes the idea of Laws of Nature is as well developed as later in Boyle and Newton. The *Discours de la Méthode* (1637) speaks of the 'laws which God has put into Nature'. The *Principia Philosophiae* (1644) concludes by saying that it has discussed 'what must follow from the mutual impact of bodies according to mechanical laws, confirmed by certain and everyday experiments'. So also in Spinoza. The *Tractatus Theologico-Politicus* (1670) distinguishes the laws 'depending on the necessity of Nature' from laws resulting from human decrees. Moreover, Spinoza agrees with Suarez that the application of the term 'law' to physical things is based on a metaphor—though for different reasons, since Spinoza was a pantheist who could not have believed in the naïve picture of a celestial lawgiver.

Zilsel sees one essential component in the development of seventeenth-century Laws of Nature in the empirical technologies of the sixteenth century. He points out that the superior craftsmen of that time, the artists and military engineers (of whom Leonardo da Vinci was the supreme example), were accustomed not only to experimentation, but also to expressing their results in empirical rules and quantitative terms. He instances the small book *Quesiti ed Inventioni* of Tartaglia (1546) in which quite exact quantitative rules were given for the elevation of guns in relation to ballistics. 'These quantitative rules of the artisans of early capitalism are, though they are never called so, the forerunners of modern physical laws.' They rose to science in Galileo.

Here the most fundamental problem is why, after so many centuries of existence as a theological commonplace in European civilization, the idea of Laws of Nature attained a position of such importance in the sixteenth and seventeenth centuries. It is, of course, only a part of the whole problem of the rise of modern science at that time. How was it, asks Zilsel, that in the modern period the idea of God's reign over the world shifted from the exceptions in Nature (the comets and monsters which had disturbed medieval equanimity) to the unvarying rules? His answer, which must surely be in principle the right one, is that since the idea of a reign over the world had originated from a hypostatization into the divine realm of men's conceptions of earthly rulers and their reigns, we should look at concomitant social developments to reach an understanding of the change which now took place. And indeed with the decline and disappearance of feudalism and the rise of the capitalist state there occurred a disintegration of the power of the lords and a great increase in the power of centralized royal authority. We are familiar with this process in Tudor England and eighteenth-century France; and while Descartes was writing, the English Commonwealth had taken the process even further towards an authority which was centralized but no longer royal. If, then, we may relate the rise of the Stoic doctrine of Universal Law to the period of the rise of the great monarchies after Alexander the Great, we may find it equally reasonable to relate the rise of the concept of Laws of Nature at the Renaissance to the appearance of royal absolutism at the end of feudalism and the beginning of capitalism. 'It is not a mere chance', says Zilsel, 'that the Cartesian idea of God as the legislator of the universe developed only forty years after Jean Bodin's theory of sovereignty'. Thus the idea, which had originated in a milieu of 'oriental despotism', was preserved in rudimentary form through two thousand years, to awake to new life in early capitalist absolutism.

Zilsel's interpretation is illuminated rather strikingly by the fact (newly revealed in the work of Crombie) that the expression 'Laws of Nature' was clearly used by Roger Bacon (A.D. 1214 to 1292), but that in the thirteenth century it did not 'catch on'. Roger Bacon wrote, for instance: 'that the laws of reflection and refraction are common to all natural actions I have shown in the treatise on geometry (*que vero sint leges reflexionum et refractionum communes omnibus actionibus naturalibus, ostendi in tractatu geometrie*)'. Vision, he said, must come about in such a way 'that it does not transgress the laws which Nature keeps in the bodies of the world (*ut non excedat leges quas Natura servat in corporibus mundi*)'. Yet he believed the power of the soul to be able to override these laws, for he added that in the twisting nerves it 'made the species (of the thing seen) relinquish the common laws of Nature and behave in a way that suits its operations (*unde virtus anime facit speciem relinquere leges communes Nature, et incedere secundum quod expedit operationibus ejus*).' This might be interpreted as a highly sophisticated idea, if Roger Bacon was really trying to say that the processes within living organisms obey higher laws than those sufficient for the inorganic world. But however he meant it, the idea of laws of matter and light simply did not win general acceptance in his day, and th concept remained dormant until at the Renaissance a new political absolutism and a new birth of experimental science brought it again into the limelight of discourse.

For the present purpose, then, it suffices to say that between the time of Galen, Ulpian, and the Theodosian Constitution on the one hand, and that of Kepler and Boyle on the other, the conceptions of a natural law common to all men, and of a body of Laws of Nature common to all non-human things, had become completely differentiated. With this established we are in a position to see in what way the development of Chinese thought on natural law and the Laws of Nature differed from that of Europe.

The ancient Taoist thinkers (Tao Chia, fourth and third centuries B.C.), profound and inspired though they were, failed, perhaps because of their intense mistrust of the powers of reason and logic, to develop anything resembling the idea of Laws of Nature. With their appreciation of relativism and the subtlety and immensity of the universe, they were groping after an Einsteinian world-picture without having laid the foundations for a Newtonian one. By that path science could not develop. It was not that the Tao, the cosmic order in all things, did not work according to measure and rule, but the tendency of the Taoists was to regard it as, for the theoretical intellect, inscrutable. It would perhaps not be going too far to say that this was the reason why, when to them was consigned the care of Chinese science through the centuries, that science had to develop on a mainly empirical level. Moreover, it is not irrelevant that their social ideals had less use than those of any other school for positive law; seeking to go back to primitive tribal collectivism, where nothing was formulated and written down, but everything worked well in communal co-operativeness, they could not have been interested in the abstract law of any lawgivers.

The Mohists (Mo Chia), on the other hand, the followers of Mo Ti, together with the Logicians (Ming Chia), strove mightily to perfect logical processes, and made the beginnings of applying them to zoological classification and to the elements of mechanics and optics. We do not know why this scientific movement failed, perhaps it was because che Mohists' interest in Nature was too strongly bound up with their practical aims in military technology;

at any rate these schools ceased to exist after the upheavals of the first unification of the empire (230 B.C.). They seem to have approached no nearer than the Taoists to the idea of Laws of Nature. The proper translation of their technical term *fa* (identical with 'law' as used by the Legalists) in the logic of the *Mo Ching* (Mohist Canon) is a very debatable matter, but so far as can be seen the conclusion that the term was used by the Mohists in a sense fairly closely resembling the Aristotelian causes, holds good.

With the Legalists (Fa Chia) and Confucians we are in the realm of pure sociological interest, for neither of these schools had any curiosity about Nature outside and surrounding man. The Legalists laid all their emphasis on positive law (*fa*), which was to be the pure will of the law-giver, irrespective of the generally accepted morality, and capable of being quite contrary to it if the welfare of the State should so require. But the law of the Legalists was at any rate precisely and abstractly formulated. As against this the Confucians (Ju Chia) adhered to the body of ancient custom, usage, and ceremonial, which included all those practices, such as filial piety, which unnumbered generations of the Chinese people had instinctively felt to be right—this was *li^a*, and we may equate it with natural law. Moreover, it was necessary that this right behaviour be taught, rather than enforced, by paternalistic magistrates. Confucius had said that if the people were given laws and levelled by punishments, they would try to avoid the punishments but have no sense of shame; but that if they were 'led by virtue' they would spontaneously avoid disputes and crimes. The *Li Chi* (Record of Rites) speaks, in symbolism appropriately taken from hydraulic engineering, of good customs as dykes or embankments, saying that while it is easy to know what has already happened it is difficult to know what is going to happen. Good customs, therefore, more flexible than formulated laws, prevent disturbances before they arise, while laws can only operate after they have arisen. Hence one can understand the point of view which after the victory of the Confucians over the Legalists came to dominate Chinese thinking, that since correct behaviour in accordance with *li^a* always depended on the circumstances, such as the status of the acting parties in social relationships, to publish laws beforehand which could take insufficient account of the complexity of concrete circumstances was an absurdity. Hence the severe restriction of codified law to purely criminal provisions.

We have dwelt already upon the distinction between li^a and fa. Neither of these words was easily applicable to non-human Nature. But there was one ancient Chinese word which does seem to link the spheres of non-human phenomena and human law. This word is *lü*. In the Chinese legal codes, it stands for 'statutes' and 'regulations'. This sense is undoubtedly quite old, as the phrase in *Kuan Tzu* may witness: 'the laws serve to distinguish each person's portion and place, and to put a stop to quarrels'. Here the idea is very close to that of μοῖρα and the other Greek entities discussed by Cornford. But the word had also a quite different meaning, namely the series of standard bamboo pitch-pipes used in ancient music and acoustics, and the twelve semi-tones which these pipes represented. What could have been the connection between the laws of sound and the laws of human lawgivers?

The word *lü* has as its right-hand phonetic a sign which was certainly in the most ancient times a hand holding a writing-implement, and for its radical the word *chhih* which meant a step with the left foot (paralleling *chhu*, a step with the right foot). This suggests an original connection with the notation of a ritual dance. Later on, since the twelve semi-tones were made to correspond with the months of the year, the word came to mean a calendrical date, and thus is found associated with the word li^e in titles of chapters on calendrical science, such as the Lü Li Chih of the *Chhien Han Shu*. The question at issue is how the conception of laws, statutes, or regulations can have been derived from, or even associated with, the word for the standard musical tones.

Perhaps the etymological considerations just mentioned hold one clue. It would not be so far a step from the directions for music and ritual dancing laid down by a diviner or priest-magician (indeed a shamanistic *wu*) to the directions for conduct of other behaviour, especially organized military behaviour, laid down by a temporal ruler. There was a logical analogy between what dancing would do against the spirits and what drilling and weapon-practising would do against human enemies. Some kinds of dances certainly involved the carrying and brandishing of weapons (Granet). It is thought that originally there were five stations around the dancing-floor which in time gave their names to a certain quality of sound, according to the instrument stationed in each place, and later to a difference in pitch.

A general connection is obvious between the musical notes on the one hand and the directions for ritual dancing and for military activity on the other. But there is nothing here which suggests that the Chinese ever thought of the semi-tone intervals of the stand-ard pitch-pipes as originating from, or constituting, any kind of law in the non-human phenomenal world. The fact that what we now regard as a branch of physics stood at the origin of a word which took on the sense of human legal regulations, may have several probable explanations, and does not, in short, mean that Chinese thinking therefore here contained the elements of the conception of Laws of Nature.

If, at this stage, a reader should happen to glance at the astro-nomical chapter of the *Shih Chi* (Historical Records), written about 90 B.C., he might well come upon the following passage:

'As for me [Ssuma Chhien refers to himself], I have studied the memoirs of the historians, and have examined the movements (of the heavenly bodies). During the past hundred years it has never happened that the five planets have made their appearances without (from time to time) moving backwards, and when they move backwards they are at the full and change their colours. And moreover, there are definite times when the sun and moon are veiled or eclipsed, and when they move to the north or the south. These are *general laws*.'

(tr. Chavannes)

In the light of the whole discussion of this lecture, he will then turn to the Chinese text fairly certain that whatever Ssuma Chhien actually said, he did not speak of general laws in the sense of the scientific Laws of Nature. Now the actual expression he used is *tu; tzhu chhi ta tu yeh*, and this word therefore demands notice.

The primary meaning of *tu* is 'degree of measurement', and that this is overwhelmingly its commonest use appears not only from the lexicographers but also from the indexes or concordances which have been made for many of the most important ancient Chinese books. Its etymology, such as might be deduced from oracle-bone forms, does not throw any light on how it came to mean this. Nevertheless, its significance may be that of 'law' especially when it is found in combinations such as *chih tu*, 'government *tu*', or *fa tu* 'legal *tu*'. Couvreur gives examples of these uses from the *I Ching* (Book of Changes), where the former combination occurs, and from the *Shu Ching* (Historical Classic), where *tu* occurs alone in the sense that certain people had 'gone beyond the bounds' or 'transgressed'. There is of course a close semantic connection between 'law' and 'measure', for every law has a certain quantitative aspect; 'How far', we say, 'is it true that such-and-such an action comes under the scope of such-and-such a provision of the law?' — 'Measures must be taken, by means of by-laws, to curb such-and-such a practice which is growing up.' But this quantitative aspect tends to remain metaphorical until legislators set out to make positive law, independent of morality, as for instance when the first Emperor Chhin Shih Huang Ti began to regulate the gauge of chariot-wheels. Still, there are to be found, among the writings of the philosophers of the Warring States period, numerous analogies between law in human societies and the carpenter's square, the compasses, and the plumb-line.

More important is the fact, pointed out by Couvreur, that *tu* may be considered a definite technical term for the movements of the heavenly bodies. The word was used throughout Chinese history for the 365¼ degrees into which the celestial sphere was divided, and for many other scales of divisions, such as the 100 parts of a day or night as shown by the clepsydra (water-clock). Revealing is the phrase used by Tung Chung-Shu in his *Chhun Chhiu Fan Lu* of about the same time as Ssuma Chhien, where he says *Thien Tao yu tu*; the Tao of Heaven has its regular measured movements. The general conclusion to which we must come is that on the strictest standards of the philosophy of science Chavannes was not justified in translating the word *tu*, standing alone, as 'general laws'. It would have been preferable to say: 'These phenomena all have their regular measured (or measurable) recurrent movements.'

One wishes that it were possible to ask of Ssuma Chhien the question: In using the word *tu*, measured degrees, did you mean it to have the undertone of 'law'? If so, whose law? I believe that it is exceedingly unlikely that he would reply 'The Laws of Shang Ti' (the ruler above); and almost certain that he would say it was *tzu-jan tu*, natural measured movement, or *Thien Tao tu*, the movements of the Tao of Heaven.

Still keeping within the realm of ancient Chinese astronomical and cosmological thought, there is to be found, in an obscure work of early date, part only of which has come down to us, a discussion very much to our purpose. This is the *Chi Ni Tzu* book, contained now in the famous collection of fragments made by Ma Kuo-Han. We do not even know whether Master Chi Ni was a real person, or simply a character invented by whoever it was who wrote the *Chi Jan* chapters or book attributed to Fan Li. Fan Li himself was a historical person, a statesman of the southern State of Yüeh in the fifth century B.C., but from the internal evidence, the discussions which Master Chi Ni (Chi Yen), carried on with King Kou Chien of Yüeh can hardly have been written before the time of Tsou Yen (late fourth century B.C.). It is of course possible that part at least of these chapters might be a Han fabrication, but it has to be admitted that they contain rather archaic material, such as the names of the spirits or legendary ministers of the five elements, and in view of their origin they may perhaps be placed in the late fourth or early third century B.C., and considered to embody a southern tradition of naturalism. The interesting lists of plants and minerals which they contain therefore class them among the oldest Chinese scientific documents which have come down to us. In any case their exact date and provenance do not affect the present argument.

In the Nei Ching section (which also survived in the *Yüeh Chüeh Shu*) we find the following:

'The King of Yueh said: "Since you discuss human affairs so brilliantly and advocate careful consideration before action, perhaps you can tell me whether natural phenomena have maleficent or auspicious meanings (in relation to man)?"

Chi Ni answered: "They certainly have. It is the Yin and Yang within the myriad things that gives them all their *chi-kang* (i.e. their fixed compositions and motions with regard to other things in the web of Nature's relationships). What fortune and misfortune depend upon are the cyclical movements of the sun, moon, stars and planets, and the recurrent alternations of destruction and generation (in the seasons of the year). For (the *chhi* of the elements) Metal, Wood, Water, Fire and Earth dominate alternately (in their own long-term rotation), and the (influence of the) moon in its waxing and waning is especially strong (at regularly recurring times). Yet all these changes are but (fluctuations) in the fundamental cyclical regularity (of the Yin and Yang in the Great Tao), which has no master (or governor, to whom, for instance, one could pray). If you follow it you will get prosperity, if you go against it you will meet with misfortune. Thus the sage (ruler) can clearly predict (the coming of) destruction and make preparations to counterbalance it, for by taking advantage of the time of lush growth he may avoid the injury of the horns of adversity. All affairs must in fact be managed following the (movements of the) Yin and Yang as exhibited in the four seasons. If these principles are not carefully used, human affairs will get into trouble. The people's livelihood is too important to allow of schemes that deviate from wise action. If you want to try to change the regularity (of the Great Tao) and the numbers (of which the world is constructed), you will only encourage unnatural acts, fall into poverty and shorten your life. Thus the sage (ruler) rejects the temptations to which little men succumb, and operates quietly as befits him, hoping to influence the unenlightened. But the mass of men strive after wealth and honours, utterly ignorant of the balance (of Yin and Yang, which will determine their fate)."

The King said, "Excellent".'   (tr. auct.)

In this remarkable and profound passage the abnormal is divested of all supernatural quality and shown to be part of a greater normal. The thought is truly advanced for its time in that extreme statistical fluctuations are regarded as fully natural departures from the ordinary, however dreadful they may be, and not as 'acts of God.' Droughts and floods, diseases or locust plagues, though coming at seemingly irregular times and posing great problems for man and society, have long-term recurrences which could in principle be predicted and against which the wise ruler

317

will protect himself and his people as far as possible beforehand.[1]
Here it would have been only too easy for the unwary to translate
chi-kang as Laws of Nature. Forke cautiously used the words
'bestimmte Wandlungen', fixed changes. Yet the lexicographers
admit the meaning of human laws in some sense for this expression.

It is obvious etymologically that we have to deal here with an
analogy from textiles; both words have the silk radical. *Chi*
combines 'silk' with 'self', it comes from an uncertain bone graph
and means 'to disentangle silk threads one from the other, to put in
order, to regulate, rule, law, norm, regular series, cycle of years,
conjunction of the sun and moon, inscribed annals'. We know
that the most prominent of such cycles of years is the Jupiter
cycle, and significantly *Chi Ni Tzu* speaks about this, giving it as
twelve years, elsewhere in the fragment. *Kang* combines 'silk' with
'net' and the ancient graph shows for the phonetic a net and a
man. From its orignal meaning of the cord forming the selvedge
of a net it came to mean 'rule, regulate, dispose, put in order,
direct', especially when used with *chi*. The analogous word
wang, though restricted more closely to the meaning of 'net',
came to imply punishments, and hence law, perhaps because of
its analogical use in the *Tao Tê Ching* (Canon of the Virtue of the
Tao). It was on the basis of these undertones that the translation
of the expression *chi-kang* in the above quotation was first adopted.

It is striking that a number of the interpretations of the words
in question imply an active verb, to disentangle, to set in order, to
rule, to make(?) laws. They derive from the oldest recorded use of
the expression in one of the *Shih Ching* odes (perhaps eighth century B.C.) where it is said that the king gives the *kang-chi* to the
four quarters of the kingdom, i.e. that he ordains its constitution
and customs. We ought not to think of this too strictly in terms of
positive law, however, for the *Shuo Wên* dictionary of A.D. 121
often mentions 'the three kang and the six chi', and an entire

[1] Western parallels for this kind of thinking concern the theory of perturbations.
For Aristotle, developmental abnormalities in animals were 'natural but not according to nature'. Through the commentators Simplicius and Philoponus the idea of
partial abnormality caused by external factors and subsumed into overall normality
passed into classical mechanics.

chapter in the *Pai Hu Thung Tê Lun* (Comprehensive Discussions at the White Tiger Lodge) of A.D. 80 explains all these as being the unbreakable threads and filaments of relationship in human society, for example those between prince and minister, father and son, husband and wife, etc. We are thus once again in the presence of Chinese natural law, and indeed *kang-chi* occurs frequently in Han texts as a legal term with this significance. If then the kings of old promulgated it, they were only acknowledging the power of something far greater than themselves, the very Tao of human society, not imposing their arbitrary will on the four quarters which they governed. And when we turn back to the world of non-human nature the position is just the same. The *Chi Ni Tzu* book expressly disclaims the idea of a supernatural 'disentangler' or a supra-personal lawgiver. It says that the great fluctuations of Nature, however catastrophic they may be if men are unprepared for them, are only part of the normal course of the Yin and Yang in the Tao of all things; everything is in motion, but it had no need of a Setter in Motion. This Tao is in fact spontaneous and uncreated, no celestial king holding sway over it who could be moved by prayers and supplications. Look to your dykes, the king is told, gather up grain into granaries against the day of adversity, waste not the people's livelihood, and study to penetrate as deeply as may be into the ways and works of Nature so as to foresee what may come. So may man in society achieve freedom from the bondage of his environment.

Definitions and explanations of the expression *kang-chi* or *chi-kang* as applied to non-human Nature are further to be found in the ancient Chinese medical literature, with which the *Chi Ni Tzu* text has a strangely close connection (not hitherto noticed); and the medical texts with their elaborate commentaries confirm and extend the interpretation to which etymological considerations had led us. In the *Chi Ni Tzu* book itself, a few pages earlier, after describing how the legendary emperor Huang Ti assigned to their tasks the tutelary and assistant spirits of the five quarters (north, south, east, west and centre), the writer goes on to say: 'Thus all together the five directions (with their corresponding *chhi*, elements and planets) constitute the *kang-chi'*. This is nothing less than the dynamic pattern of the universe. Indeed the conception of a net is obviously very close to that of a vast pattern. There is a web of relationships throughout the universe, the nodes of which are things and events. Nobody wove it, but if you interfere with its texture you do so at your peril. In the following pages we shall be able to trace the later developments of this web woven by no weaver, this Universal Pattern, until we reach, with the Chinese, something approaching a developed philosophy of organism.

Such conceptions are taken as a matter of course in the medical classics. Thus the *Huang Ti Nei Ching Su Wên* (The Yellow Emperor's Manual of Corporeal Medicine; the Pure Questions and Answers) says: 'The Yin and the Yang constitute the Tao of Heaven and Earth and the *kang-chi* of the myriad things, the father and mother of change and transformation, the beginning and end of life and death, and the source of the mysterious movements of light and darkness'. And in another place it says: 'Thus the motions of the heavens and the quiescence of the earth (i.e. the coming and going of the Yin and Yang) is the *kang-chi* of the mystery of the universe.' The commentators of Thang and Ming give their own various further understandings; thus Ma Shih says: 'The Yin and Yang in all things as they aggregate and come into being is *kang*, the Yin and Yang in all things as they disperse (and die) is *chi*'; and Chang Chieh-Pin: 'The Yin and Yang constitute the Tao of Heaven and Earth; their sum is called *kang* and their cycle of recurrence is called *chi*'. Therefore once again we are not dealing with the laws of any lawgiver, but with the fixed compositions and motions of all particular things with regard to other things in the woven pattern of Nature's relationships.

So far, then, we have not found in Chinese thought any clear evidence of the idea of law in the sense of the natural sciences. Still keeping to the schools which considered themselves Confucian, we must turn to the Neo-Confucians of the Sung dynasty (twelfth century A.D.). Chu Hsi, and the other thinkers of his group, made a great effort to bring all Nature and Man into one philosophical system, and the principal concepts with which they worked were *li^b* and *chhi*. The latter corresponded approximately to matter, or rather to matter and energy, and the former was not far removed from the Taoist conception of the Tao as the Order of Nature, though the Neo-Confucians also used the term *tao* in a slightly different and technical sense; *li^b* could be described as the ordering and organizing principle in the cosmos. For this word Bruce, Henke, Warren, and more recently Bodde, all adopted the translation 'law', but in my judgment they were not justified in so doing, and in view of the great confusion which it is liable to cause, this interpretation should be abandoned.

The word *li^b*, in its most ancient meaning, signified the pattern in things, the markings in jade or the fibres in muscle; as a verb it meant to cut things according to their natural grain or divisions. Thence it acquired the common dictionary meaning, 'principle'. It undoubtedly always conserved the undertone of 'pattern', and Chu Hsi himself confirms this, saying:

'*Li^b* is like a piece of thread with its strands, or like this bamboo basket. Pointing to its rows of bamboo strips, the philosopher said: One strip goes this way; and pointing to another strip: Another strip goes that way. It is also like the grain in the bamboo—on the straight it is of one kind, and on the transverse it is of another kind. So also the mind possesses numerous principles (*li^b*).'

(tr. Bruce)

Li^b, then, is rather the order and pattern in Nature, not formulated Law. But it is not pattern thought of as something dead, like a mosaic; it is dynamic pattern as embodied in all living things, and in human relationsips and in the highest human values. Such dynamic pattern can only be expressed by the term 'organism', and Neo-Confucian philosophy was in fact a scheme of thought striving to be a philosophy of organism.

We seem to be in presence, then, in the latter part of the twelfth century, of a point of view rather similar to that which Ulpian had expressed in Europe nearly a millenium before, and which had been incorporated into the Justinian *Digest*. But the profound difference is that while Ulpian had spoken quite uncompromisingly of *law*, Chu Hsi relies chiefly on a technical term the primary

L

meaning of which is *pattern*. For Ulpian (as for the Stoics) all things were 'citizens' subject to a universal law; for Chu Hsi all things were elements of a universal pattern. On the whole, it does not appear that it is possible to find more than traces of the concept of Laws of Nature in the greatest of Chinese philosophical schools, the Neo-Confucians of the Sung. Their emphasis was on something different, though no less important ultimately for the natural sciences.

Another word which it has often been tempting to translate as 'Laws of Nature' is *tsê*. In the official biography of the great astronomer Chang Hêng (A.D. 78 to 139) we read: 'The steps of heaven (i.e. the number of degrees passed through by planets and constellations in a given time, their risings and settings, etc.) follow unvarying rules (*chhang tsê*)'. But one can also find instances of doubt whether it is possible for man to understand the *tsê* which operate within the things of Nature. The first I will give is from a passage in that part of the *Chhu Tzhu* (Elegies of Chhu) which contains the poetical writings of Chia I, and therefore dates from about 170 B.C.

'Heaven and Earth are like a smelting-furnace, the forces of natural change are the workmen, the Yin and the Yang are the fuel, and the myriad things are the metal. Now it runs together, now it disperses, sometimes moving and sometimes resting. Effortless and natural are these processes, but do they really have fixed rules (*an yu chhang tsê*)? In the thousand changes and the myriad transformations there is never any final end nor any absolute beginning.'

(tr. Forke, mod. auct.)

The second is from the commentary of Wang Pi on the *I Ching* (Book of Changes), and must therefore date from the close neighbourhood of A.D. 240. Explaining the twentieth hexagram symbol, *kuan*, meaning 'view', or 'vision', he says:

'The general meaning of the Tao of *kuan* is that one should not govern by means of punishments and legal pressure, but by looking forth one should exert one's influence (by example), so as to change all things. Spiritual rule is without form and invisible (*Shen tsê wu hsing chê yeh*). We do not see Heaven command the four seasons, and yet they do not swerve from their course. So also we do not see the sage ordering the people about, and yet they obey and spontaneously serve him.'

(tr. auct.)

This is perhaps the most illuminating passage of all. We have a flat denial of the conception of orders issued to the four seasons (and hence the courses of the stars and planets) by some celestial lawgiver. The thought is extremely Chinese. Universal harmony comes about not by the celestial fiat of some King of Kings, but by the spontaneous co-operation of all beings in the universe brought about by their following the internal necessities of their own natures. *Tsê* is really the internal rule of existence embodied in each individual thing, whereby it conforms to its position and function within the whole of which it is a part. One begins to see how deeply rooted in ancient Chinese ideas was the Neo-Confucian philosophy of organism. In Whitehead's idiom, the 'atoms do not blindly run' as mechanical materialism supposed, nor are all entities specifically directed on their paths by divine intervention, as spiritualistic philosophies have supposed; but rather all entities at all levels behave in accordance with their position in the greater patterns (organisms) of which they are parts. Thus *tsê* never meant anything like the Laws of Nature in the Newtonian sense, and such an interpretation cannot properly explain the thought about *li*[b] of the Neo-Confucians.

The affirmation that Heaven does not command the processes of Nature to follow their regular courses is linked with that root idea of Chinese thought, *wu wei*, non-action, or unforced action. The legislation of a celestial lawgiver would be *wei*, a forcing of things to obedience, involving imposition of sanctions. Nature shows a ceaselessness and regularity, yes, but it is not a commanded ceaselessness and regularity. The Tao of Heaven is a *chhang Tao*, the Order of Nature is an unvarying order, as Hsün Chhing says (*c.* 240 B.C.), but that is not the same thing as affirming that anyone ordered it to be so.

Thus in the *Li Chi* (Record of Rites) there is an apocryphal conversation between Confucius and Duke Ai of Lu. The Duke asked what was the most valuable thing to note about the ways of Heaven.

'The Master replied: "The most important thing about it is its ceaselessness. The sun and moon follow each other round from east to west without ceasing:

such is the Tao of Heaven. Time goes on without interruption; such is the Tao of Heaven. Without any action being taken, all things come to their completion; such is the Tao of Heaven." '.

(tr. Forke)

Here again, then, is a denial, if an implicit one, of any heavenly creation or legislation. It should be noted, in passing, that although the concept of *wu wei* was emphasized particularly by the Taoists, it was part of the common ground of all ancient Chinese systems of thought, including that of the Confucians.

It may be worth while following this profound idea a little further. It is not at all difficult to find passages which confirm the conception of Heaven acting according to *wu wei*; it runs throughout the *Tao Tê Ching*, where we find the significant statement that though the Tao produces, feeds, and clothes the myriad things, it does not lord it over them, and asks nothing of them. The idea is, in fact, a Taoist commonplace, and appears in such books as the *Wên Tzu* and many later writings. The *Lü Shih Chhun Chhiu* (Master Lü's Spring and Autumn Annals, *c.* 240 B.C.) affords us a little further insight into the working methods of the Tao of Heaven. There we read:

'The operations of Heaven are profoundly mysterious. It has water-levels for levelling, but it does not use them; it has plumb-lines for setting things upright, but it does not use them. It works in deep stillness . . .

Thus it is said, Heaven has no form and yet the myriad things are brought to perfection. It is like the most impalpable of featureless essences, and yet the myriad changes are all brought about by it. So also the sage is busied about nothing, and yet the thousand executives of State are effective in the highest degree.

This may be called the untaught teaching, and the wordless edict.'

(tr. auct.)

Such a conception is truly sublime. But it is profoundly incompatible with the conception of a celestial lawgiver. The movements of the celestial bodies proceeded, in the one case, according to teachings which no one had ever taught, and according to edicts which no one had ever issued or even put into words. But the Laws of Nature which Kepler, Descartes, Boyle, and Newton believed that they were revealing to the human mind (the very word 'revealing' is symptomatic of the spontaneous background of occidental thought) were edicts which had been issued by a supra-personal, supra-rational being. The fact that this was later generally recognized to be a metaphor does not mean that it may not have had great heuristic value at the beginning of modern science in Europe.

I conclude, therefore, that 'law' was understood in a White-headian organismic sense by the Neo-Confucian School. While one could not say that 'law' in the Newtonian sense was completely absent from the minds of Chu Hsi and the Neo-Confucians in their definition of li^b, it played a relatively minor (perhaps very minor) part, and the main component was 'pattern', and hence pattern living and dynamic to the highest extent, and therefore 'organism'. In this philosophy of organism all things in the universe are included: Heaven, Earth, and Man have the same li^b.

In Europe natural law may be said to have helped the growth of natural science because of its universality. But in China, since natural law was never thought of as law, and took a special name, li^a, it was very hard to think of it as in any way applicable outside human society, although it was relatively much more important than natural law in Europe. When order and system and pattern were visualized as running through the whole of Nature, it was not as li^a but as the *tao* of the Taoists or the li^b of the Neo-Confucians, both of which were rather inscrutable, and neither of which had a juristic content.

Again, in Europe positive law may be said to have helped the growth of natural science because of its precise formulation, and because it encouraged the idea that to the earthly lawgiver there corresponded in heaven a celestial one, whose writ ran wherever there were material things. In order to believe in the rational intelligibility of Nature, the Western mind had to presuppose (or found it very convenient to presuppose) the existence of a Supreme Being who, himself rational, had put it there.

This brings us back to the Taoists. The Taoists, though profoundly interested in Nature, distrusted reason and logic. The

Mohists and the Logicians fully believed in reason and logic; but if they were interested in Nature it was only for practical reasons. The Legalists and Confucians were not interested in Nature at all. Now this gulf between empirical nature-observers and rationalist thinkers is not found to anything like the same extent in European history, and as Whitehead has suggested, this was perhaps because European thought was so dominated by the idea of a supreme creator being, whose own rationality was the guarantee of rational intelligibility in his creation. Whatever may be the needs of mankind now, such a supreme God had inevitably to be personal. This we do not find in Chinese thought. Even the present-day Chinese translation of Laws of Nature is *tzu-jan fa*, 'spontaneous law', a phrase which uncompromisingly retains the ancient Taoist denial of a personal God, and yet is almost a contradiction in terms.

Here we cannot investigate the ancient Chinese conceptions of God. An immense literature exists on the subject, since the Christian missionaries in the last few centuries engaged in great debates as to the correct translation of European terms; most of this is now not worth the paper on which it is written, since at that time sinological studies were in their infancy. We know that the most ancient terms for God in Chinese were *Thien* (Heaven) or *Shang Ti* (The Ruler Above) though other terms were used, for example *Tsai* (Governor) in *Chuang Tzu*. *Thien* is undoubtedly an anthropomorphic graph (presumably of a deity) in its most ancient form, and *Ti*, though not absolutely clear, is distinctly anthropomorphic also. So, I think, is *Tsai*, which is related to a character meaning 'demon'. Much sinological work is being done on the extent to which there was in ancient China a personalization of these conceptions, and it is hard to summarize such conclusions as have been reached. Many theories are in the field: Creel, for instance, thinks that *Shang Ti* was a transcendentalization of the function of the emperor (or bronze-age High King); Granet considers that he was a personification of the calendrical order of the seasons; another view, represented by Fitzgerald, looks upon him, and upon *Thien*, as symbols of the Original Ancestor. Creel expresses the now generally received opinion that *Shang Ti* is the older of the two, being associated with the Shang dynasty, while *Thien* is a rather later Chou dynasty term. Tai Kuan-I believes that the name *Shang Ti* was taken over from the Miao peoples by the Chinese. But in any case three things are clear: (a) that the highest spiritual being known and worshipped in ancient China was not a Creator in the sense of the Hebrews and the Greeks; (b) that the idea of the supreme god as a person in ancient Chinese thought, however far it went, did not include the conception of a divine celestial law-giver imposing ordinances on non-human Nature; and (c) that the concept of the supreme deity very early became quite impersonal. It was not that there was no order in Nature for the Chinese, but rather that it was not an order ordained by a rational personal being, and hence there was no guarantee that other rational personal beings would be able to spell out in their own earthly languages the pre-existing divine code of laws which he had previously formulated. There was no confidence that the code of Nature's laws could be unveiled and read, because there was no assurance that a divine being, even more rational than ourselves, had ever formulated such a code capable of being read. One feels, indeed, that the Taoists, for example, would have scorned such an idea as being too naïve to be adequate to the subtlety and complexity of the universe as they intuited it.

To sum up, therefore, I would say that the conception of the Laws of Nature did not develop from Chinese conceptions of law in general, for the following reasons. First, the Chinese acquired a great distaste for precisely formulated, abstract, codified law from their bad experience with the School of Legalists during the period of transition from feudalism to bureaucratism. Secondly, when the system of bureaucratism was definitely set up, the old conceptions of *li*[a] proved more suitable than any others for Chinese society in its typical form, and therefore the element of natural law was much more important relatively in Chinese than in European society. But the fact that so much of it was not put in formal legal terms, and that it was overwhelmingly human and ethical in content, made it impossible to extend its sphere of influence to any form of non-human Nature. Thirdly, the available ideas of a supreme being, though certainly present from the earliest times, became depersonalized so early and were so lacking in ideas of creativity that they prevented the development of the conception of precisely formulated abstract laws ordained from the beginning by a celestial lawgiver for non-human Nature, and capable, because of his rationality, of being deciphered or re-formulated by other lesser rational beings using the methods of observation, experiment, hypothesis, and mathematical reasoning.

The Chinese world-view depended upon a totally different line of thought. The harmonious co-operation of all beings arose, not from the orders of a superior authority external to themselves, but from the fact that they were all parts in a hierarchy of wholes forming a cosmic pattern, and what they obeyed were the internal dictates of their own natures. Modern science and the philosophy of organism, with its integrative levels, has come back to this wisdom, fortified by our new understanding of cosmic, biological, and social evolution; though who shall say that the Newtonian phase was not an essential one? And lastly, always in the background, stood the concrete forces of the social and economic life of Chinese society, out of which arose the transition from feudalism to bureaucratism, and which could not but condition at every step the science and philosophy of the Chinese people. Had these conditions been basically favourable to science, the inhibiting factors considered in this lecture would perhaps have been overcome. But all we can say of that science of Nature which then would have been developed is that it would have been profoundly organic and non-mechanical.

Before concluding, we may glance at a striking illustration of the difference in outlook between China and Europe in the matter of nature and law. It is generally known that during the European middle ages there was a considerable number of trials and criminal prosecutions of animals in courts of law, followed frequently by capital punishment in due form. Scholars have gone to the trouble of collecting a large amount of information on these cases. Their frequency follows a curve with a well-marked peak at the sixteenth century, rising from three instances in the ninth to about sixty in the sixteenth, and falling to nine in the nineteenth century; and it seems doubtful whether this is due, as Evans suggests, to lack of adequate records for the earlier periods. The peak corresponds to the witch-mania (Withington). The legal actions fall into three types: (a) the trial and execution of domestic animals for attacking human beings (for example the execution of pigs for devouring infants); (b) the excommunication, or rather anathematization, of plagues or pests of birds or insects; (c) the condemnation of *lusus naturae*, for example the laying of eggs by cocks. It is the last two which are most interesting for the present theme. In 1474 a cock was sentenced to be burnt alive for the 'heinous and unnatural crime' of laying an egg, at Basel; and there was another Swiss prosecution of the same kind as late as 1730. One of the reasons for the alarm involved was perhaps that *oeuf coquatri* was thought to be an ingredient in witches' ointments, and that the basilisk or cockatrice, a particularly venomous animal, hatched from it. But the interest of the story lies in the fact that such trials would have been absolutely impossible in China. The Chinese were not so presumptuous as to suppose that they knew sufficiently well the laws laid down by God for non-human things to obey, to enable them to indict an animal at law for transgressing them. On the contrary, the Chinese reaction would undoubtedly have been to treat these rare and frightening phenomena as *chhien kao* (reprimands from heaven), and it was the emperor or the provincial governor whose position would have been endangered, not the cock. Let us quote chapter and verse. In the long Wu Hsing Chih (Discussion of the Five Elements) in the *Chhien Han Shu* (History of the Former Han Dynasty) there are several references to sex-reversals in poultry and in man. These were classified under the heading of 'caerulean misfortune' (*chhing hsiang*) and thought of as connected with the activities of the element Wood. They forboded serious harm to the rulers in whose dominions they occurred.

As regards the second of the three types of prosecution mentioned above, it is interesting that the European medieval attitude wavered. Sometimes the field-mice or locusts were considered to be breaking God's laws, and therefore subject to prosecution and conviction by man, while at other times the view prevailed that they had been sent to admonish men to repentance and amendment.

It is extremely interesting that modern science, in so far as since the time of Laplace it has been found possible and even desirable to dispense completely with the hypothesis of a God as the basis of the Laws of Nature, has returned, in a sense, to the Taoist outlook. This is what accounts for the strangely modern ring in so much of the writing of that great school. But historically the question remains whether natural science could ever have reached its present state of development without passing through a 'theological' stage.

In the outlook of modern science there is, of course, no residue of the notions of command and duty in the 'Laws' of Nature. They are now thought of as statistical regularities, valid only in given times and places, descriptions not prescriptions, as Karl Pearson put it in a famous chapter. The exact degree of subjectivity in the formulations of scientific law has been hotly debated during the whole period from Mach to Eddington, and such questions cannot be followed further here. The problem is whether the recognition of such statistical regularities and their mathematical expression could have been reached by any other road than that which Western science actually travelled. Was perhaps the state of mind in which an egg-laying cock could be prosecuted at law necessary in a culture which should later have the property of producing a Kepler?

NOTE

Three Chinese terms in the preceding lecture are homophones and have therefore been distinguished by superscript letters. They are:

li[a] good customs, ceremonial observances, ethical behaviour.

li[b] organic pattern at all levels in the cosmos.

li[c] calendrical science based on observational astronomy.

CHRONOLOGY OF CHINA

		B.C.
	HSIA kingdom (legendary?)	*c.* 2000/*c.* 1500
	SHANG (YIN) kingdom	*c.* 1500/*c.* 1030
Chou Dynasty	Early Chou period	*c.* 1030/722
(Feudal Age)	Chhun Chhiu period	722/480
	Warring States period	480/221
First	CHHIN dynasty	221/207
unification		

		A.D.
	HAN dynasty	
	Earlier or Western Han	202 B.C./A.D. 9
	Hsin interregnum	9/23
	Later or Eastern Han	25/220
	Three Kingdoms period (San Kuo)	
First	Shu (west)	221/264
partition	Wei (north)	220/265
	Wu (south-east)	222/277
	CHIN dynasty	
Second	Western	265/317
unification	Eastern	317/420
	Former (or Liu) SUNG dynasty	420/479
	Northern Wei dynasty (Tho-Pa Tartar)	
	later split into Eastern and Western	386/554
	Northern and Southern empires	
	(Nan Pei Chhao)	479/581
Second	Chhi (southern)	479/502
partition	LIANG	502/557
	Chhen	557/581
	Chhi (northern)	550/581
	Chou (northern)	557/581
Third	SUI dynasty	581/618
unification	THANG dynasty	618/906

	Five Dynasty period (Wu Tai)	907/960
	Later Liang	907/923
Third	Later Thang	923/936
partition	Later Chin	936/946
	Later Han	947/950
	Later Chou	951/960
	LIAO dynasty (Chhi-Tan Mongol)	907/1125
	Hsi-Hsia State	990/1227
Fourth unification	Northern SUNG dynasty	960/1126
Fourth partition	Southern SUNG dynasty	1127/1279
	CHIN (Ju-Chen Tartar) dynasty	1115/1234
Fifth unification	YUAN (Mongol) dynasty	1260/1368
	MING dynasty	1368/1644
	CHHING (Manchu) dynasty	1644/1911
	Republic	1912

INDEX
by M. Moyle

335

INDEX

INDEX

Index